HIMALAYAN ECOLOGY

Encyclopaedia of Environmental Pollution and Awareness in 21st Century Series

HIMALAYAN ECOLOGY

Edited by

V.K. Prabhakar

ANMOL PUBLICATIONS PVT. LTD.

NEW DELHI - 110 002 (INDIA)

ANMOL PUBLICATIONS PVT. LTD.
4374/4B, Ansari Road, Daryaganj
New Delhi - 110 002

Himalayan Ecology

First Edition, 2001

ISBN 81-261-0908-4

PRINTED IN INDIA

Published by J.L. Kumar for Anmol Publications Pvt. Ltd., New Delhi - 110 002 and Printed at Mehra Offset Press, Delhi.

Contents

PREFACE

Today, the cry of 'pollution' is heard from all the nooks and corners of the world, and pollution has become a major threat to the very existence of mankind on this earth. It is the major challenge of our times.

The pollution of various resources has gone to such an extent that we are unable to breathe fresh air and drink fresh water. On one hand the advancements of science and technology have added to the human comforts by giving us automobiles, electrical appliances, supersonic jets, space crafts, better medicines, better chemicals to control harmful insects and other pests, etc., but, on the other hand, they have given us a very serious problem to face that is pollution. Pollution can be defined as an undesirable change in the physical, chemical, or biological characteristics of the air, water, or land that can harmfully affect health, survival, or activities of humans or other living organisms.

Pollution does not have to cause physical harm, but it may merely interfere with human activities. Chemicals spewed into the air or water from an industrial plant may be harmful to humans and other organisms living near the plant. If expensive pollution controls are required, the plant may be forced to shut down.

India faces serious problems of environmental pollution. Water pollution from untreated community and industrial wastes flowing into our rivers and streams, increasing use of fertilizers and pesticides, dumping of organic wastes etc., are issues that need to be tackled on the priority. Air pollution, long regarded as the bane of industrialised nations looms large over our country.

Here, we are presenting this *Encyclopaedia of Environmental Pollution and Awareness in Twenty-first Century* in fifty volumes with a view to make people conscious about the seriousness of pollution in the present time and its remedies.

Editor

1 INTRODUCTION

The Himalayan range stretching from Pakistan to Arunachal Pradesh in India must be one of the most spectacular and dramatic areas on earth. Not only does it embrace the highest mountains in the world, but also mighty glaciers and ferocious rivers, huge foothills, and fertile valleys, extensive deserts and dense forests. And it is not only the topography that is so impressive, there is a considerable variety of mountain people with different cultures, religions and language. Four decades ago tourism, as we know it today, was almost non-existent. And possibly that has been the major change in the Himalayan activity—the development of number and quality of airfields in the mountains.

Nowadays tens of thousands of people are walking the hill tracks, floating down the rivers, driving over the great Himalayan passes, landing in tiny airfields in remote valleys, visiting the sacred places in the mountains. The impact of all these people visiting the distant mountain areas has been tremendous. Some of the effects are good and some not; so good.

The standard of living of the mountain people has increased considerably from tourist wealth. But the beautiful tracks and valleys have been littered with rubbish and the forests have been destroyed to provide fires for meals and comfort for tourists. Hundreds of tiny hotels do the landscape but toilets are rare and completely inadequate. The Himalayan environment has suffered severly, partly from tourism but more importantly from the substantial growth in the local population. Arable land is increasingly rare and the forests are cleaned and the slopes terraced for the planting of crops.

Population growth is a massive problem to overcome—nobody really knows how to deal with it. It is only in affluent countries that the population growth is stable. In the Third World a large family is the only method of social security and there is a constant and desperate race between industries.As the tourist numbers grow, the impact on the most popular areas becomes more and more severe. But in the huge stretch of the Himalayas there are a multitude of beautiful areas and many of these are completely neglected. Energetic efforts should be made to attract people to the less popular places, to spread tourism thinner, so that their negative environmental impact is reduced.

Tourism agencies must be pressurised into ensuring that their customers leave the hills and valleys in as pristine a condition as they were when they found them. One of the great joys of adventure tourism is getting away from the everyday whirl of living, appreciating the peace and contentment of areas, free from public pressure, enjoying the bubbling streams, quiet forest paths, and lonely mountain spaces. We all need to escape from the concrete jungle and return refreshed to the city environment and dream of sparkling waters and icy mountain tops. The Himalayas are good for all of us but we must not let them be abused.

Forests must be preserved, camp-sites must be kept clean and wholesome, and rivers protected from pollution. We must prevent the wanton destruction of all that is fresh and beautiful and enjoy to the full, the superb beauty of nature in the remote mountain areas.

The environmental hazards in the Himalayas are many, complex, and interrelated e.g. landslides, droughts, floods, decline in the forest cover, decline in agricultural yield, soil loss, problem of siltation in reservoirs and rivers, extinction of rare species of wild animals, etc. The natural springs are rapidly dying, domestic animals are less productive or unproductive, the fodder resources are limited and natural forests are fast disappearing in the Himalayas. Blasting of the rocks for the construction of roads, tunnels, dams, colonies. etc and haphazard method of mining have greatly affected the ecological balance. Mining operations are old and traditional and have erred the lush green vegetation cover.

The mining operations have also eroded the lush green hill slopes, destroyed leaf mould—a valuable organic fertiliser, polluted streams, and decreased the agricultural production by 60%. In the doon Himalayas, the limestone dust particles from the quarries and lime factories have been estimated at 360 micrograms per cubic meter as against the safety level of 100. Hazardous air and noise pollution have caused many plants to wither away; the forest under-growth has been cleared, the sub-surface water has decreased, and the mountain side resembles a gigantic skeleton due to blasting.

In higher altitudes, frost edging is an environmental hazard. The glaciers do a lot of abrading and scratching and sometime snow streams and avalanches cause a great loss to life and ecology. In areas of moderate relief of the Himalaya, water erosion is a serious problem as it makes the rock naked; consequently, instability of the hill slopes, silting of the river valleys and floods are very common features. Torrential rains cause mass wasting and huge landslides and landslips. Landslides are very common, often blocking the vehicular traffic and sometimes damaging engineering projects. In the low relief alluvial plains, flash floods, soil erosion and sheet erosion are environmental hazards. The construction of unimaginatively planned and cruelly executed roads is the biggest factor in destroying the ecological balance. Construction of dams on hill rivers to meet the ever increasing demand of irrigation water and power of the plains is yet another environmental hazard. These projects cause destruction of hills, flood the fertile flatland of the valleys, and create the problems of rehabilitation of the oustees.

The Himalayas are the loftiest and greatest of all the mountain system in the world. They have from time immemorial held an indescribable charm, inspiration and spiritual significance for the millions living in and around these mountains. It is well impossible to describe in words the mysticism and the religious aura that surround these mountains. Great Sages and poets have tried to describe the sheer majesty and charm of the Himalayas but could never fully capture in words their majestic; grandeur.

The Himalayan land and water are priceless national assets. The approach of the modern man to explore and exploit its resources is very harsh. If the Himalayan region to meet the increasing demand for fuel, fodder, timber, etc., would be disturbed and it would create environmental hazards in the Himalayas. The question of conservation, regeneration and development cannot be left to chance and the vagaries of political and administrative set-up at the lowest level. Conservation and development should be made mandatory without prejudice to the rights which inhere in the land. In this way, objections or dilatory tactics can be given short shrift.

The Himalayas are an awesome chain of mountains that extend approximately to 2500 km. From north-west to south-east and the width is between 250to 300 km. The

Himalayan system consists of three longitudinal belts separated by faults, their geological and topographical characters are strikingly different.

Immediately north of the plains is the Siwalik Zone. The width of this zone varies from 10 to 50 km. and altitude rarely exceeds 900m. This zone is covered with damp and unhealthy forest. The next zone is known as the middle Himalaya with width ranging from 65 to 75km. and the average height being 300m. This zone is profusely dissected by the major rivers passing across it. The third belt is the Great Himalaya rising to an average height of 5,200m. The great Himalayan peaks are situated in this zone. There are more than 90 peaks having an altitude of more than 8,000m. This zone has two sub-zones, viz, an alpine zone (4800m.) and an upper snow-bound zone (7500m).

The transverse sub-divisions of the Himalaya are as under:

(i) The Punjab Himalaya—Sutlej to Indus, 590 km.

(ii) The Kumaon Himalaya—Kali to Sutlej, 350km.

(iii) The Nepal Himalaya—Tista to Kali, 800km.

(iv) The Assam Himalaya—Brahmaputra to Tista, 750km.

Dr. Valdiya, an eminent geologist has strongly supported the rise of the Himalaya out of a vast tethys geosyncline that existed where now we see the Himalayas. The soft sediments that lay at the bottom of the geosyncline were pressed, upheaved in several phases. The present configuration and complexity of the structure of the Himalayas is due to the phenomenon of "plate tectonics" and other related orogenic activities.

The Himalayas are very young mountains and constitute less than 1/5th of the total geographical area of India. They control the climate and through the rivers the prosperity of more than one third of the total population living in the plains of the Indus, the Ganges and the Brahmaputra.

The Himalayas as the name implies are the "abode of snow". The total ice cover of the Himalayas is around; 16,000sq.km. and this amounts to 17% of their area. A very large number of glaciers exist at present above the snow line, most of which are in the Himalayan range. Glaciers provide water to major perennial rivers of the Himalayan system. Glaciers are formed from the accumulated snow which has survived melting; for a glacier to exist, the snowfall should exceed the quantity that melts in a summer. There are around 15,000 glaciers in the Himalayas. They vary in length and majority of them are one to five kilometers long.

The Gangetic glacier in the Central Himalaya is the source of the Ganges. Its ice surface occupies little over 20 cu.km. of ice. Satopath is another glacier which forms the secondary source of the Ganges. The Brahmaputra which ranks among the mighty rivers of the sun-continent of India, rises in the Great Glacier in the Kailash range in the northern most chain of the Himalaya just south of Lake Tasho in the Tibetan region. The Indus, the Ganges and the Brahmaputra are the three major river systems having their source in the Himalayas. They drain the country and provide a large potential for the irrigation and domestic use and also for the generation of hydroelectric power.

In implementing the development programmes it is necessary to ensure that the adverse environmental impacts do not arise. The total catchment area of these three rivers is about 8,00,000sq.km. Out of the total utilizable surface water resources of 666 milliard cu.m. Over 245 milliard cu.m.is accounted by these three mighty Himalayan rivers. Indirectly, the Himalayas play a vital role in replenishing the water resources of India.

They act as a barrier for moisture laden monsoons and enable to precipitate their moisture on the southern slopes. The melting of the snow and enormous rain keep the drain age system of the Himalayas healthy and further enrich the subsurface flow of water. The Himalayan glaciers are the banks of water. Harnessing of the Himalayan water resources should be taken up in such a manner that ecosystem is not disturbed. The productivity of water resources can be enhanced by trout culture in the free floating water in higher altitude areas and by raising mahasir and mirror carp in the standing waters with extensive use of cage culture technology.

Based on practice, there are two distinct agricultural systems in the Himalayas, i.e., "shifting agriculture" and "sedentary agriculture". Shifting agriculture is being done in the eastern Himalayas while sedentary agriculture is very popular in the western part of the Himalayas. Out of the total area (under cultivation), about 80% is under seasonal crops. The area under tree crop is around 12%. The area under tree crops is shrinking, giving way to season bound crops. Land under the crop in the higher Himalayas has shown an increase of about 15%. Double cropped area in the Himalayas was about 33% on an average, varying from 41% in the Siwaliks to 20% in the middle Himalayas. Sector wise, it is 38% in the central Himalaya and 10% in the Western and the Eastern Sectors. Cultivated land holdings in the Himalayas are smaller as compared to the holdings in the plains. In the outer Himalayas there are two distinct cropping seasons, i.e., Rabi and Kharif.

In the Rabi crops are included wheat and barley which are planted in autumn and harvested in early summer. In the Kharif crops are included paddy, maize, etc. which are planted in spring and harvested in late autumn. Agriculture in the Western and Central Himalayas require hard work for at least ten months in a year. In the Eastern Himalayas, it is an year round affair. Among the crops, rice cultivation starts from lower altitudes and extends upto an altitude of 1800m. Altitudinally rice is followed by maize and wheat which are grown in slopes above the rice fields. At still higher levels, millets and barley take over. The Himalayas are the major supliers of temperate fruits, which do well in between 1200 to 2000m. However, the gross area cropped under fruits is very small.

Agriculture in inaccessible regions of the Himalayas is difficult and requires considerable hard work. Terracing cultivation is common, even in the steepest slopes. On an average terrace is 3 to 4m long, 2 to 2m wide and 2 to 3m high. Shifting cultivation or Jhumming practised in the north east has very deep roots and very wide ramifications as the tribal groups system of cultural and ethnic types is typical of the area and has resited change for a very long time. The practice is a low energy budget, a low investment system of subsistence husbandry which uses the comparatively rapid mineralisation and recycling of nutrients provided there are sufficient areas available. Now that there is not enough area for the system to function, degradation is rampant.

About 0.542 million hectare area is under shifting cultivation and about 2.64 million tribal population is affected by it. It is now accepted that this practice has to be rationalised. The limited arable land should be utilised scientifically for valuable cash crops which offer a regional and seasonal advantage and for the development of seed industry for temperate vegetables which cannot be produced in hill zones and adjacent plains. Different cash crops need to be determined based on agroclimated suitability.

Horticulture, for which even slopy marginal lands

can be exploited, hold much greater promise than food grains. Instead of encouraging only one product (apple), a mix high value low bulk non-perishable fruit cultivation need to be encouraged and popularised in the Himalayas. The traditional agricultural practices are generally faulty; these should be discouraged systematically and ecologically viable scientific techniques adopted. The low products agrarian land should be brought under agro-forestry, farm forestry and agri-horticulture. Regarding forestry, the forests must be developed beyond the traditional forestry and a significant amount of total demands in terms of fuel, fodder, timber and other tree products can be met from the wastelands and crop fields, if integrated programmes such as social or farm forestry can be implemented at community level considering the needs of inhabitants and site characteristics.

The existing forest land should not be alienated for other land use, particularly agriculture and horticulture. The degraded forest should be enriched by the species already existing in such biotopes as these are well acclimatised to the locational factors compared to the exotic species. Free range grazing as practised in the region is injurious for regeneration of forest. Local inhabitants should be educated about the merits of stall feeding and rotational grazing. Afforestation programme should be encouraged; and popularised as major component of the rehabilitation and conservation strategy. There should be adequate stock of seedlings and seeds to maintain an ecological balance, it is necessary to build up productive assets and to increase the supply of consumer goods, fuel, food, fibre, fodder, and products of resin, etc.

The life and environment in Himalaya is a symbiotic relationship between agriculture, forestry and animal husbandry. The Jersey cow has been found as the best milch animal for Himachal Pradesh while buffalo for Central Himalaya. The local livestock can be improved; by cross breeding the unproductive animals should be replaced. The Angora rabbits may be introduced as an alternative to local sheep for wool to particular extent. The bovine population may be reduced by castration of local bulls. The existing mode of open feeding should be stopped. The poultries to be encouraged as considerable income is generated through these.

Silvipastoral or Silvo-hortpastoral systems in the Himalayas are more useful and profitable compared to grass development alone. Effective protection of grazing land would increase the herbage cover and herbage yield. Fertilisation and introduction of grass legume mixtures in mid hills and cocksfoot, timothy, personal rye grass and contucky grass in sub-Alpine and Alpine grasslands can regenerate the degraded grass resources. Introduction of plantation of fodder trees on terraces and private lands shall improve the position of green top feed during the lean periods. Thus, planting improved grass varieties or strains coupled; with more efficient management of grasslands is vital in meeting current and future needs of herbage providing protective soil cover.

The Himalayas offer tourists a variety of scenic beauty. To them it is a paradise, its gushing rivers are laden with trout and the varieties of fish. To trekkers and hikers it is a challenge, its canyoins, deep valleys and escarpments are eye-catcher. The snow-peaked mountains, gliding glaciers, grass meadows and hot springs invite tourists from remote corners of the world. The ever increasing tourist influx is creating great impact on ecology and environment. The problem of litter, noise pollution and damage to natural assets is causing great anxiety to environmentalists.

The unmanaged influx of tourists destroys the

natural architecture of the rocks and natural surroundings. The development and construction of new hotels involves the destruction of vegetation, felling of trees and erosion of the environment. Tourism should be considered as on alloy and not an adversary in conserving the ecology in the Himalayas. It should be in the fitness of the things if master plans are prepared and new areas are develped to meet the unending demand of tourists resorts. There should be a close coordination among the planners and the promoters of ecology and environment of the Himalayas.

The construction of roads and other infrastructure are indispensable for efficient transport and communication system for tourism development, but these unavoidable activities are mainly responsible for destruction of Himalayan land resource along its vegetation and animal wealth. A careful study shows that the construction of 44,000 km. long roads in Himalayan region produced 2,650 million cubic metres of debris and each kilometer of already constructed roads generates; 550 cubic metres of debris by landslides and rockfalls.

Thus nearly 24 million cubic metres of debris slide down damaging vegetation and chocking the springs etc. These activities accelerate hill slope instability deforestation, soil erosion, pollution of water and air, etc, especially along the road sides and around the most of the Himalayan tourist resort like Nainital, Simla, and Mussoorie etc. The use of powerful explosives like dynamite and frequent heavy traffic create vibrations which casue cracks on rocks and earth which in turn lead to land slides and erosion along with adverse impact of the flora and fauna. The construction of roads etc. along geologically active zones like at Kaliasur, Satpuli(Garhwal) is also responsible for degration of land. The construction for accomodation facilities like rest houses, tourist bunglows, buildings and seasonal accomodation units, recreation centres,etc., have posed the most serious menace to the land resources of the region; not only does it accelerate soil erosion but it also causes damage to crops, animals and plants, housing and public properties. It is also a contributory factor in causing floods with heavy siltation in valleys and adjacent areas. In Chamoli district alone from 1971 to 1978, 48 human lives, 949 heads of cattle, 601 houses, costing nearly about 173 lakhs of rupees, as also the crops, were destroyed by mass movements.

In the year 1972 owing to the landslides, the Gohna Lake was formed and its breaching has resulted in a large scale destruction in the Alaknanda valley and the plains of U.P. The flash flood engulfed 2000 sq.k.m. paddy crop, 10k.m. roads, 6 motor bridges and 24 buses, 366 houses while 138 villages were effected; by landslides alone in the year 1979.

Agriculture is one of the main occupations of the Himalayan inhabitants, but it does not give quick and heavy economic gains like tourism. At present during the tourist season (mainly May-June) tourism has become the main source of income, as a result of which agricultural system has miserably regressed. Enormous agricultural land is being occupied for development of infrastructure facilities. These activities create dependence system in addition to adverse impacts on economy and ecology of the area. These activities create dependence system in addition to adverse impacts on economy and ecology of the area.

The farmers who are living in the vicinity of these tourist resorts leave their traditional farming and either work or conduct different types of business for prompt and easy economic benefits. In many such developing and well developed tourist places, tourism has almost overlapped agriculture. Fields around the famous tourist places like Mussorie, Nainital, Simla, Darjeeling, etc.are the result of this type of tourism impact.

The beauty of the Himalayan forest, alpine meadows and peaks is the main sources of tourist attraction. The Continual flow of tourists, mountaineers, trekkers, etc. is likely to have positive as well as negative impacts on these natural resources to a certain extent. Mythological beliefs and tradition which are prevalent in the Himalayan region are also responsible for adverse effect on the flora, specially in those localities adjoining the shrines. Religious tourists habitually collect flowers and colurful herbs to offer to temples like Gangotri, Yamnotri, Badrinath, and Kedarnath, etc. Consequently, many species which are part of our cultural and religious heritage are beginning to get wiped out. It was noticed by the authors during the Roopkund trekking in Chamoli district that at the Bedney Bugayal and Roopkund (places of mythological importance along the trek) thousands of Brahmakamal flowers (Saussurea obvollata) were offered which were plucked from vicinity area.

A recent sample study of Surkanda Devi (a place about 25km. from Mussorie towards Tehri), which is situated at a most picturesque spot in the highest (2750m) peak of outer Himalayan range, indicates that about 10,000 people gather there for a week on the occasion of Dashera (celebrated festival of Hindu) and practically every one offers few flowers with twigs of Ronsli (Abies Pindrow) to Goddess Surkanda along with their other gifts. Every visitor roughly plucks ten flowers and two twigs. In this way about 10,000 flowers and 20,000 twigs are lost annually, as a result of which a virgin Abies pindrow forest and densely grown different colourful flowers which we have seen during childhood (about 23 years back) have nearly disappered now, and in their place scrub forest and flowers like Primula denticulata, Anemone vitifolia, etc. are sparsely found. In addition to this, severe trampling by visitors is also reponsible for damage. A recent field survey of the world famous valley of flowers also shows that its flora and fauna was under major attack and influenced adversely after 1975 when this area was opened to visitors. Due to increasing rate of tourism traffic and direct (like trampling and plucking etc.) and indirect activities of tourists, many important and charming species of flowers like blue-scented Primulas (most beautiful), red Potentilla spp., creamy Anemone spp., blue Meconopsis spp., etc. are now found in reduced abundance along the treks, road sides and other easily accessible areas. This degradation will definitely occur in the entire area with increasing number of tourists.

In addition to this, other factors like overgrazing, low grazing etc. are also responsible for ecological imbalance. Overgrazing prevents plants from flowering and fruiting willow or ungrazing accelerate the growth of unnecessary weeds and grasses which create hinderence in growth and blooming. The growth of unwanted grass and weed has become a major problem of the valley after the ban imposed on grazing by the government and it seems to be only one reason for low flowers occurence in the valley even in peak season (August-September) in comparison with the situation obtaining some years ago. This fact leads to the conclusion that a scientific and controlled grass cutting or grazing is necessary for maintaining the importance of the valley of flowers.

If the sensitive ecosystem of the valley will not be given proper importance and attentions along with the promotion of tourism, Smythe's silent valley of Garhwal Himalaya may one day cease to show its visual spectrum, of nature's beauty. The damage in the Himalayan environment has reached a critical level and beyond the carrying capacity; therefore, in order to overcome this problem, it is the need of time that certain more landscapes with unique type of vegetation and wildlife should be conserved as "Biosphere Reserves".

The tourist concentration in the Himalayan alpine pastures is increasing day by day, more than the carrying capacity, while the area under vegetation is decreasing due to accelerated human activities. A considerable area of virgin forest is also being replaced for providing necessary amenities to visitors for tourism promotion. It has been estimated that 234.25 sq.km. forest area has been destroyed only by road construction from 1951 to 1979 in the U.P. hills (forest statistics 1981). Deforestation not only result into resource depletion but also deteriorate fresh and underground water, wildlife and other related resources along with acceleration of disastrous phenomena like heavy soil; erosion and landslides etc., which mainly provoke flash floods.

The forest provides the best shelter and source of food for most of the wild animals, hence the quality of forest affected by vast tourism directly influences the wildlife of the respected area. The ever increasing influx of tourists exerts an adverse impact on the unique Himalayan wildlife. Tourist activities like invasion in wildlife habitat, close view watching and photography, hunting, poaching, and other disturbances etc. have created a problem for wild animals.

Mammals and birds are affected in large extent—more than reptiles, insects and other small animals. Owing to the large interference of man in the habitat and life style of animals specially in national parks, wild life sancturies etc., the ferocity of animals is being lost which is the only quality of wildness that differentiate them from domestic animals. The act of providing bate to tiger or lion for close view in national park etc., spoil the quality of wild life and is destruction of wild animals in the name of tourism. These activities further affect the feeding and breeding habits and ultimately the entire life cycle. The small animals die due to heavy and frequent traffic which affect the food chain in wild life system. The habit of tourists to provide ready made food to wild animals is also responsible for loss of ferocity. During the year 1975 in the Corbett National Park a wild boar popularly known as Buddha (innocent) had become a purely domestic one.

Owing to the availability of ready made food like bread, biscuits, chapati, etc. from tourists he had later on developed the habit of stealing edible items like lunch packets, chocolates, jam butter, even tinned meat and fish, sauce and drink bottles, etc., from tourist tents, kitchen, or even open car, jeep etc. thereby a purely herbivourous wild animal had changed his wild habits. Recently in the year 1987 is has been observed by the authors in Corbett National Park that young chittal (spotted deer) is also frequently visiting the Dhikala complex of the park to eat cooked food material like bread, etc., from the hand of tourists and officials.

It is also noticed that during night time; few wild animals like wild boar (Susscrofa), fox (Vulpeseulpes), chittal or spotted deer (Axis) etc., reach close to the Dhikala complex and feed on thrown-away ready made food or garbage etc. sometime even together also. So it is necessary to protect these animal habits and habitat alongwith their natural ferocity to maintain the wildness and ecosystem balance of natural food chain system between predator and the prey or between the herbivorous and herbs for total ecological system.

The crowded gathering of people religious places and resorts bring marked change to ecosystem as a whole. The direct and side products of tourists like litter, polythene bags and wrappers, garbage piles and related materials attract rodents and crow, which in turn draw bigger animals and birds, thus changing the entire habit and habitat. The increasing tourist's craze for the Himalayan wild life and their products is also responsible for its destruction as it increases the demand of stuffed animals, ivory ornaments, furs, tails, quills, feathers, horns, skins, musk and wild-life souvenirs, etc. This demand provokes the curio trade, resulting into poaching and slaughtering of animals.

The musk deer, (Moschus moschiferus), blackbear (Selenaretos thivetanus), snow leopard (Panthera unica), leopard (P. pardus), bharal (Pseudois nayar) monal (Lophophorus impejanus), etc. are few well known animal species which are in great danger due to poaching. To prevent or reduce degradation of the Himalayan natural resources, there should be scientific and well planned growth of tourism in which experts of different disciplines like geology, botany, geography wildlife, environment, etc., and technocrats like engineers, architect etc. along with local people should have close involment in planning. There should be recreation between natural resources and tourism development which should be according to the carrying capacity of particular resorts, so that space action does not affect their ecosystems and ecological balance. If these protective measures are not given due importance in tourism planning and development, the Himalaya would be damaged beyond the limit of repair of regeneration.

Industrial development in the Himalayas is slow but the rapid population explosion and urbanizations has changed the trend. In the Kashmir Himalaya, the Himachal Himalaya, the Kumaon Himalaya, the Assam Himalaya and in many other parts of the Himalayas fruit processing, wool processing cottage industries, cement industry, paper and pulp industries, resin processing and other forestbased industries, such as watch factories, tool factories, television factories etc., with pollution control devices would be vital for the rapid economic growth of the Himalayan belt.

The Himalayas provides a variety of natural resources to the sub-continent of India, including the life-giving water. These resources have been threatened largely by human negligence and activities. In order to improve and regenerate degraded hill environment, cientific approach involving hazard zone mapping and banning engineering development activities in such zones, drainage control and dewatering measures, protection of slopes through vegetation, modification slopes where necessary, application of engineering technology and improved agronomic practices should be carried on priority basis. The ecological balance can be maintained and improved by reforestation, by checking the illegal felling of tree, controlling the course of floods by construction bounds, retaining walls, schecking landslides, rocketries, sinking of the soils, protecting the watersheds, etc.

Laws should be enacted to stop indiscriminate timber trade, installation of new cement factories, stone crushers, state factories, industries, etc, which in turn cause air, water and noise pollution. The mining operations should be done using modern methods and techniques to minimize the pollution of atmosphere, the ruthless and haphazard mining activities greatly affect the health of the workers and also create ecological imbalance by rock debris, dumps and quarries' wastes.

The greenery of the soils is lost and fertility of lands is decreased. Constructional and developmental works also destroy the natural setting of the scenery and landscape. Soil and land use inventories need to be taken up on priority basis and the Himalayan belt may be divided into viable seacoast for the purpose of land use, planning and management. Modern techniques of remote sensing and aerial photo interpretations are very vital ion conserving the soil and soil programme. Soil maps may be employed to carry out the afforestation programmes and proper landuse. Aerial seeding should be carried out to reclaim the barren lands. To prevent or reduce degradation of the Himalayan natural resources, there should be scientific and well planned growth of tourism in which experts of different disciplines like geology, botany, geography, wildlife, environment, etc and technocrats like engineers, architect etc. along with local people should have close involvement in planning.

There should be correlation between natural resources and tourism development which should be according to the carrying capacity of particular resorts, so that the space action does not affect their eco-system and ecological balance. If these protective measures are not given due importance in tourism planning and development, the prosperous heritage and environment of the Himalaya would be damaged beyond the repair of regeneration. To protect our environment we have to protect our wild-life, rare birds and unique plants. Proper education on environmental rehabilitation and conservation must be initiated at the grass root level. Rehabilitation and conservation programmes should include such measures as finding alternative sources of energy, self employment, agriculture, sericulture, horticulture, social forestry, cottage industries, small scale industries, etc.

The human and animal husbandry programmes should also be taken up to improve the lot of the Indian village folks. Women are the backbone of human ecosystem in the Himalayan hills. It is, therefore, necessary to improve their lot. An all-out effort should be made to maintain the enchanting beauty of the Himalaya.

Environmental Management of The Himalayas

It is only during the last few decades that man has begun to exploit the vast and varied natural resources. The great rivers which originate from the springs and the snowfields possess immense potential for a multiplicity of uses. At several places these rivers have been tapped and plugged to generate hydroelectric power, to construct dams and generate power, to irrigate the land, and to transport the material through the rivers. A proper management of these rivers and other natural resources to benefit the inhabitants of the holy mountains is very vital. The problems of environmental management are diverse and require special attention.

The Himalayas extend from the west to the east for about 2,500 km. and they very in width from 250km. to 300km. These mountains have the world's highest peaks. The diverse landscape soils and climatic sectors, support several types of vegetation. Some of the plants are very characteristic of tropical, sub-tropical, cool temperate, temperate and alpine conditions. Some of the regions also support desert or semi-desert vegetation.

The Himalayan ranges can be divided into three major botanical regions i.e. the western, the central and the eastern Himalayas. Of these, the western region is further subdivided into the western Himalayas, Himachal Pradesh and the north western Himalaya. The central Himalayas more or less, conform to the mountainous region in Nepal. The eastern Himalayas includes Sikkim, Bhutan, Arunachal Pradesh, etc. Considering that the Himalayan region is richly endowed with a wide range of plants and is, in fact, one of the centers of origin of certain important agricultural and economic plants, an earnest need is to survey plant wealth, especially rich in genetic diversity.

The survey, however, should not be merely a

taxonomic one but a dynamic one carried out by multidisciplinary experts with the object of conserving the rare species and those species which are vital in the field of medicine, forestry, agriculture. etc. An effort should also be made to check them growing wild. The exotic plants, e.g. rhododendron, wild roses, coniferous trees, and rare herbs and food crops should be genetically upgraded to meet the requirement of food.

The temperate Himalayan belt offers great potentials for forage grasses and legumes, as such the natural grasslands of low productivity can be replaced with highly productive by farming systems. These untapped grassland resources of the Himalayan region can perhaps make the largest contribution to its economic development. As the prosperity of a nation is chiefly determined by its capacity to utilize and conserve its resources, it is therefore, necessary to undertake the steps based on the physical and biological factors of the region so as to formulate a plan for the rapid development, economic progress and sound management of the Himalayan plants.

The wild mammalian fauna of the Himalayas shows a tridimensional zonation, i.e. (a) Longitudinal, (b) Latitudinal and (c) Altitudinal. In each major zone, number of subzones can be recognized. They show interesting adaptations and extralimital affinities. The Himalayan mammals are provided with long and thick hairy covering atleast in winter to cope with the severe cold. Several animals resort to migration to lower heights during winter.

Animals confined to the coniferouns regions show a darker colouration as on adaptation to increased humidity and also as a protection against ultraviolet radiation, but the animals capable of living above the forest zone tend to be paler: e.g. the snow leopard, the

Horsfield's leopard cat, the Trevelyan's leopard cat, the pillar's cat, the snow bear, the Himalayan ermine and pikas. The member of Artiodactyla which have speed as the chief means of escape from the enemy are well adapted for quick movements on the steep and rocky slopes of the mountains by means of their splayed and pointed hooves.

Some carnivores like tiger, lion, panther, cat, etc., show highly interesting adaptations of stalking the prey on the difficult terrain of the high altitudes. In the case of the snow leopard, the ears are very small and the eyes are placed high up on the head so that the animal is able to reconnoitre for prey and locate it with a minimum possible exposure of head. In the case of the pillar cat, the short ears are set very wide apart and low down on the sides of the head so that their inner and upper borders hardly rise above the forehead which is highly flattened. The adaptation is of vital advantage for stalking the prey without the exposure of the predator from behind the rocks.

The wildlife in the Himalayas has been greatly affected by rapid development and urbanization. The rapid deforestation has also resulted in the extinction of wildlife. The wholesale degradation of the Himalayas, indiscriminate poaching of wildlife for commercial, medicinal, and industrial purpose has greatly reduced the stock of the wildlife.

The illegal trade in protected animals now has higher profits than drug trafficking. A kilo of rhino horn costs about Rs. 2,50,000/- and a kilogram of musk can fetch upto Rs.7 lakhs in the international market. The number of animals killed in laboratories for academic and research work is difficult to ascertain. Cosmetics, shampoos, etc; from big industrial concerns are first tested on the eyes of rabbits resulting in excruciating pain. The trade in hides, skins and live animals is indeed cruelsome. The use of pesticides and insecticides to eradicate deleterious insects cause harmful impact on the wildlife of the Himalayas. For an effective and successful conservation of wildlife, it is necessary that deforestations, spraying of pesticides, insecticides, trade in animals and experimentation on animals should be legally stopped.

The number of parks and sanctuaries should be increased to preseve the vanishing rare species. The proper management of the sanctuaries and national parks is very vital. They should be protected from poachers and those who trade in animals. Factors limiting their population should be tackled seriously. Among the pests are fruit bats, rodents, lagomorphs, etc. Their control, conservation and proper management can be done by the agriculturist and the silviculturist. The population of the insectivorous bats can be increased by providing roosting boxes for these highly beneficial animals. A well planned management programme is bound to conserve the fauna of the Himalayas.

The Himalayan region is an intricate ecological complex stretched over an area of 61 million hectares of which an ared of 17.8 million hectare is under forests and another of 0.7 million hectares in covered by alpine pastures. The ecosystem ranges from a high humid thick jungle belt in Assam to the dry sub-zero desert wastelands of Spiti and Ladakh.

The entire Himalayan belt can be sub-divided, on the basis of bioclimatology, into three belts i.e. the dry cold deserts of Ladakh and Spiti, the true temperate hills, and the humid and the high rainfall hills. Consistent with this ecological pattern, the livestock resources have developed, having become both dependent on man and independent of him. The Himalayan belt supports 20 million cattle, 10 million buffaloes, 2.9 million sheep, 5.94 million goats, 0.12 million horses, donkeys, etc; 1.27 million pigs. 17.38 million poultry and other birds and other livestock comprising 1.18 million heads which include yaks, charoes. etc. The animal resources are large and can be a source of proteins for human consumption, a source of fibre, leather, dairy products. meat products. etc.

A sound environment requires gradual replacement of animal resources by better stock and through the addition of new highly productive genes. A crop-livestock ecosystem mix which can maximize the yields of animal products without disturbing the environment should be adopted.

The Himalayas act as a barrier for the monsoon and cause precipitation to enable millions of people to carry out agricultural operations in the extr-peninsula and Indo-Gangetic plain. The annual rainfall varies in different ecological frames and structures from 5 cm in Ladakh to 400 cm in Arunachal Pradesh. Precipitation in the form of snow is very common in the higher atltitudes.

The soils are mostly acidic at higher altitudes in the eastern Himalaya. They are found in association with nonacidic soils at mid altitudes of the western Himalaya and at low altitudes of U.P. Himalaya. The meadow soild are also acidic in character. An acute problem with the soils is their erosion. The degree or intensity of erosion is very high and is of the nature of sheet erosion, gully erosion, etc. Owing to steep slopes, erratic landuse, indiscriminate grazing, lopping and felling of trees at random, flash floods, silting etc. the problem has become acute. Improper landuse, small and scattered holdings, limited irrigation and old (or conventional) methods of agriculture need due attention.

A more rational land use of the resources can be introduced in the region by carrying out systematic soil surveys and preparing soil and land use inventories. The demarcating lands suitable for agriculture or horticulture or both. The chopping system and crop rotation studies do the least damage to the environment while bring to the farmer maximum returns from his land. Experiments should be initiated to determine the ways and means of aligning the rows of plants and the shape, heights and contours of the terraces to minimize losses of the precious "top soil" due to runoff. etc.

Considering that domestic animals are indispensable to the people of Himalaya, a programme should be launched to develop pastures and suitable leguminous plants, and fodder grasses should be introduced in the region. Keeping in view the spiral rise in population, it is recommended to supplement the food production in the region by encouraging the fish industry and fish-culture. The trout fish is becoming very popular, the economic lot of the people living in the vicinity of the lakes, springs, ponds, etc. where the trout fish is found. There is also need for overcoming Jhuming way of cultivation in the eastern Himalaya.

The proposed management and soil conservation techniques would certainly make the region agriculturally prosperous and self-sustaining. According to an eminent scholar, Dr. M.S.Swaminathan, the regions of the Himalayas should be studied with reference to weed plants, their biology, their relationship with the crop plants, their distribution, control, diseases, etc. There are many medicinal and aromatic plants which need special investigation for making economic use of them.

Studies on orchids, several type of bulbous plants and mushrooms will definitely help to improve the economy of the people in these regions. Studies on the harvesting, storing and sale of crops is of utmost importance. The problem of rats, bamboo rats and other rodents should also be solved. The fesibility of alternative land use and constraints in the adoption of new techniques should be given priority. It is an opportune time to fight the unholy tripple alliance of pests, pathogens and weeds.

The place of forests in the Himalayan environment depends upon the vegetational and geological evolutionary history, longitude, latitude, altitude, drainage patterns, climatic patterns, and soild conditions. Their role in regulating precipitation, streamflow, subsurface flow, supplying food, fodder, fruits, etc. is vital. Their preservation and maintenance is of paramount importance to maintain the ecological balance. In order to conserve the forest resources and to check their rapid depletion, it is essential to link up forest reources with forest-based industries on the basis of physiographic, industrial and economic catchment concepts. The rural economy can be improved by several schemes of social forestry.

The wastelands and degraded lands be employed for "organized forestry," "multipurpose farm forests," etc. Keeping in view the importance of vegetation cover and sound landuse practices for soil and water conservation, it is felt that there should be a close co-ordination among the irrigation, soil conservation and forest departments both at the State level and at the Centre. This shall help to ensure optimum social and economic returns to the local inhabitants without bringing imbalance in the ecosystem.

The evolving of sound management practices should form the core of all scheme pertaining to afforestation and regeneration. Intensive study should be made to identify the suitable species of trees for wastelands and the degraded lands.

The problem of silt discharge is very acute in the Himalayan streams. An adequate network of gauging stations be established in the hill streams. These stations should be equipped with adequate instruments, equipments, etc., to monitor the qualitative and quantitative analysis of silt discharge and sedimentation throughout the year. An effort should also be made to control siltation in all catchment areas. Recognizing that the utilization of the irrigation potential of the Himalayan region is sub-optional, it is imperative to survey the irrigation potential for sound water management. Since srings are vital sources of drinking water in large parts of the Himalayas, it is, therefore, very necessary to undertake the hydrological investigations to ensure a regular utilization of the spring water.

To prevent water run off and judicious use for protective irrigation, small reservoirs to suit local conditions, soil, climate, topography, etc. should be constructed. As most of us know the Himalayan glaciers are the source of several streams, it is, therefore, necessary to make an exhaustive study of the Himalayan glaciers and also prepare an inventory of glaciers, glacial debris, glacial moraines, etc.

Managing the Physiographic and Catastrophic Hazards

The geologically young are of the Himalayas, the presence of residual stresses and highly deformed and tectonized condition of the rocks make them highly susceptible to weathering, erosion and damage by structural and seismic activity. The developmental activities in the Himalayas have to face the challenge of these geological hazards, and in some cases the developmental activities, such as the impounding of large quantities of water in reservoirs, tunnelling operations and the addition of silt load of rivers by road cutting, lead to the intensification of the natural dynamic processes. The weathering is mostly due to block disintegration and thrawing. This causes the production of vast quantity of Scree or talus. The debris or scree slips down along the slopes under the action of gravity. The debris falls on the fertile fields, stops the free circulation of water, roads are blocked and landslides of intense magnitude are caused. The Badrinath area of Uttar Pradesh presents a typical example of subaerial denudation and landslides. Intense grazing has also accelarated the processes of soil erosion and this has made the slopes barren, naked and prone to sheet erosion and gully erosion.

Enormous quantities of silt and sediments fill the river valleys and result in flash floods. The catastrophe of the river Brahmputra alone annually carries a suspended load of 800 million tonnes and the average sediment yield from its catchment in India is of the order of 26,000 hectares metres. The Siwaliks in the outer Himalaya experience mass wasting and this results in huge landslides. Such landslides are very frequent near Nashri Nala (Kud-Batote) in Jammu, near Poonch (Jammu), Malori on the Batote-Bhadarwah road of Jammu, Mussorie highway on the Chakkir-Dalhouse road and on the hilly regions of the eastern Himalaya.

The landslides have also been observed in the Kumaon and Garhwal Himalayas and they cause a great loss of life and property. Dr. K.S.Valdiya, an eminent scholar of the country, is of the opinion that the tectonic stresses which heaved up the mighty Himalayan mountains have not died down as yet, and provide an additional impetus and vigour for the dynamic processes. These processes are manifested by intense seismic activity, mass wastage, floods, transgressions, sinking, subsidence, regressions, etc. Most of the earth-quakes that have been felt in the extra peninsular region are due to the dynamic forces and till date million of people have died besides an irreparable loss to landscape and property.

The earth temors also cause huge landslides. The creep movement in the areas of Chamoli district of Uttar Pradesh in early 1968 resulted in a 40 metre high blockade across the Rishiganga river near the village of Reni(U.P). The lake thus formed was silted up by May, 1970 and the blockade was breached during July, 1970, and the floods in the Rishiganga river created environmental hazard. About 0.175$km^2$was reactivated owing to the erosion of the debris derived from a prehistoric slide by the Karamnasa Nadi—a tributary of the Alaknanda, resulting in a debris avalanche, which washed away the motor road built above and blocked a part of the course of the Alankanada river. Major slides occured also in other parts of the Himalayas causing blockades, floods, disruption of roads, etc. Avalanches are also very common in the higher altitudes of Leh, Kargil, Gurez etc. in Kashmir, Spiti in H.P., in the higher reaches of Kumaon-Garhwal and Badrinath area of U.P. These cause disruptions of telecommunication lines, electric poles, destruction of engineering structures, disruption of roads by the accumulation of snow and ice.

A proper appreciation of these dynamic processes and their scientific evaluation in time can decrease the intensity of loss, if not stop it altogether. The rock waste derived from the construction and excavations of the roads, bridges, reservoirs, tunnels, etc. should be placed at an area which may neither affect the cultivated lands nor the flow of a stream. The appropriate use of such a waste is in the construction of rock-filled and concrete dams or in road dressing. The environmental management of the Himalayan landscape is fraught with complex interrelationships between natural entities and the evolutionary processes, and between human activities, government concerns and programmes of resource management. The human interference for developmental activities in the Himalayas has to be in consonance with the environmental conditions. As the Himalayan terrain has some of the best tourist resorts, e.g. Kashmir, Nainital, Dalhousie, Mussori, Darjeeling, etc., it is essential that the developmental activities should be planned in such a way that their aesthetic appeal is not lost.

The maintenance and development of roads which play a significant role in the ecology of the hill terrains, should be given due attention and Central and State teams should work in close co-ordination for the development of roads. It is finally recommended that long term basic research into the functioning and dynamics of the Himalayan ecosystem should be initiated for an everlasting and sound environmental management of the the Himalayas. The following priorities for research are identified:

(i) The problems associated with the current massive erosion and landslides occuring in the hilly terrains of Himalaya.

(ii) The problems associated with the explosion in population, rapid urbanization and migration of masses in the mountains regions.

(iii) The problems associated with the rapid changing faunal and floral stocks in the Himalayas.

(iv) The problems associated with tourism and the adverse effects on the environment.

(v) The problems associated with ecological and environmental awareness and establishment of "Biosphere Reserves".

The intense investigation of the above referred problems would help in achieving the destined objects of an excellent and sound management of environmental problems in the Himalayas.

2 HIMALAYAN ECOLOGY

Water springs and streams—the lifeline of Himalayan communities—are dying especially during the non-monsoon dry periods. Land slides, roads and other construction activities also add to the causes for destruction activities. Government and non-government bodies have taken action to repair the degradation of the Himalaya. But these actions are not enough and have to made any dent into the problem. In spite of development, trading, education and public awareness amongst the people in the Himalaya, the help of NGOs and people's organisations is essential to save the Himalaya and develop self-reliant Himalayan communities. The people must decide for themselves as to what their priorities and programmes of work are.

A community needs a plant nursery, a cooperative set up for planting and care of trees, rejuvenation of water springs etc. There may be financial constraints in the course of executing these programmes but such problems could be overcome by determination and will of the people. Integrated land and water-use to build-up self-reliance could go a long way to meet people's needs.

People should understand the strategies being planned. The focus should be strengthening local capacities for planning and development activities. Participation by local communities should define who should participate in what kind of programmes and how to participate and implement the programmes such as voluntary, coercive, participatory, etc. Steps needed for large-scale people's participation for development promoters could be: Go to the people' Live with them; Learn from them; Serve them; Love them; Plan with them: Start with what they know; Build on what they have. These eight principles when applied are likely to give response to the promotion of a good economic development.

The above are the basic principles of people's participation. Various ministries and NGOs as well as voluntary agencies have taken several steps to implement the above principles by organizing.

1. Symposia, workshops, seminars, etc. of NGOs and successful cooperators in the Himalaya.
2. Eco-development camps by villagers, students, NGOs etc, in various parts of the Himalaya normally financed by Government departments.
3. Initiating plant nurseries, trees plantation and after-care.
4. Conducting training courses in colleges, schools, training institutions, KVKs, etc.
5. Initiating group action for water harvesting and other soil and water conservation programmes.
6. Promotion of generation through various types of fencing such as bio-fencing, stone-wall fencing, but mainly social fencing.

The energy conserving efficiency of the total community was maximum in summer. The values were less when estimated annually.

The efficiency of gross production in perennial herbaceous forest communities ranges between 2.0% and 3.5%. These figures would be doubted if the incident amount of photosynthetically active radiation is taken as a basis. In the tropical grass-lands, some earlier authors have determined the highest energy conserving efficiency values in February in completely protected and in open grazed and in August in semi-protected grasslands. On the whole, the energy conserving efficiency value for completely protected site was about four times greater than in semi-protected and open grazed site.

In the present study the energy conserving efficiency value in fenced was also about six times greater than that in grazed meadowland. The overall results indicated that 31 and 42% of the total net production disappeared respectively form fenced and grazed sites and in the rest energy remains locked in the vegetational compartments. Such studies were also conducted by Billore for the grazing lands of Ujjain; the values were more or less same, but some differences occurred which may be due to their different edapho-climatic and floristic influences.

Agriculture section deals with problems of development and management of sub-montane grass lands comprising the region of Siwaliks and the out hill ranges of Western Himalaya. It excludes the temperate and alpine grasslands. The upper limit of the tropical and subtropical regions is 2,100 m alt. which also is the transition zone from dry and moist Sal type and *Pinus roxburohii* type of forest to the oak-conifer forest zone. The altitudinal distribution of the various forest types, their ecological succession and parallelism with grassland has been discussed elsewhere.

The submontane region has been included under the agricultural zone 1, the main grass cover being *Themeda-Arundinella* type. Gupta and Nanda described the ecological succession of the Western Himalaya and the synecology of the grass covers for the Garhwal region. The topography is essentially mountaineous.

Anthropogenic influence in this zone due to dense settlements and migratory animal population from high altitudes during winters is very strong. cutting, grazing and lopping of the forest vegetation have marked effect on the land-scape. The general tendency is to clear forests and *benap* lands for agriculture.

Field terraces are made even on steep slopes. Irrigated terraces are only maintained properly with soil and water conservation measures while the unirrigated terraces where occasional cultivation is practised are at low level of management. Steep slopes provide condition for high degree of natural erosion and are the areas primarily for perennial vegetation mainly for grasslands. Plants like *Carissa spinarum, Euphorbia royleana, Cotinus Coggyria,* all immune to grazing are commonly seen. Grasslands throughout the country, except for the alpine altitudes and the extreme arid area are considered to be biotic or bio-odaphic in origin and show various stages of ecological succession depending on the nature of the soil and the extent of grazing and other biotic factor. No where, these represent the "climatic climax" in the Clementsian terminology

The climate of the region is characteristic of the tropical deciduous forest and forest of *Pinus roxburghii.* It is usually warm with heavy precipitation during the monsoon. Physiologically dry months range between 3-4 in the Siwaliks to 1-2 in the outer hills. Accordingly to Labroue *et al* these represent the typical bixeric bioclimates of the northwest India.

The tract falling in the Siwalik zone belongs to the quarternary era where the rocks are loose and soft in nature. Recent excavations have revealed that the area was once the bed of vast ocean in which were found Elephants, Rhinoceros, etc. The outer hills are comparatively of late age and were formed during the

Tertiary era. These consist of mainly quartzites, phyllites, limestone and shales. There is wide range of soils in the sub-montane and foot hills commonly classified as "Bhabar and Terai soils" and "brown forest soils".

The botanical composition of the grasslands for the Garhwal region and the Western Himalaya has been described elsewhere. *Heteropogaon* type is associated with the submontane zone subjected to cutting and burning of slash. This species contributes to 63 per cent of the botanical composition under protection. Associated grasses are *Bothriochloa pertusa, Arundinella setosa* and *Chrysopogon fulvus,* etc. It occurs as low as Terai region where tall grasses are subjected to burning. In as Terai region where tall grasses are subjected to burning. In the Bhabar region its forms a distinct community with *Eulaliopsis binata.*

The *Arundinella* type is primarily associated with the *Pinus roxburghii* type of forest where it contributes to nearly 66 per cent to the botanical composition under protection. Other main grasses associated are *Arundinella setosa, A. nepalensis, Heteropogon contortus* and *Chrysopogon fulvus*. The *Cenchrus ciliaris* type is primarily restricted to alluvial flats in Jammu region where it contributes to 62 per cent of the botanical composition. The associated grasses are *Cynodon dactylon, Bothriochloa pertusa, Saccharum spontaneum, Imperata cylindrica* and *Sporobolus marginatus*. The most important species for development and management of grasslands in the region is *Cenchrus ciliaris* in the Siwalik zone upto 500 m and *Chrysopogon fulvus* on the Siwaliks and the outer hill ranges upto 2,100 m. However, *Eulaliopsis binata* is another grass for the region with is primarily the source for fibre and paperpulp.

The Chir pine needles and the heavy forest litter in the dry zone is very much susceptible to fires which are both accidental and international. Chir pine needle being acidic in nature also do not allow luxuriant growth of grass cover. Otherwise a thick grass cover is evident both in the dry and moist sal type of forests. In addition to unrestricted grazing, this factor modifies the ecological succession to a great extent.

Main problems of development have been classified as below:

1. *Protection*: Where human settlements have agriculture as the main profession land holdings are very small and the people do not allot land for fodder. There is a high incidence of continuous grazing with the result that the perennial grasses have almost disappeared giving place to tough obnoxious weeds such as *Lantana camara, Euphorbia royleana,* etc. Soil erosion has taken place from community lands which are "Gochar lands," being everybody's property is no man's responsibility. Protection improvement and management of such lands is the major problem.
2. *High cost of development* : The grasslands with protection tend to progressive ecological succession where shrubs and weeds invade the area quickly. Weed eradication along with soil conservation measures is costly and uneconomic with the present standards of secondary productivity.
3. *People's participation in socialistic and democratic management of community lands:* In almost all the areas, deferred grazing is an immediate necessity till the grasslands improve with the desired species. Curtailment in grazing rights is very much desirable. This could be achieved only with the active participation of the villagers, since reduction in cattle population is more a social and politico-religious problem.

Socio-economic surveys in the submontane zone have shown that on an average a household keeps 6 animals comprising of 1 buffalo, 2 bullocks and 3 cows. Goats and sheep are dominant in the inner hill ranges. Only about 25 per cent of the cows are in milk and rest are dry. The milk yield is poor. On an average, a cow and a buffalo give 112 and 734 litres of milk respectively during their location period. The animals are kept for dung which is the primary source for manure in the fields are hardly any farmer supplement the FYM with fertilizers.

Animal husbandry occupies the leisure time of the farmer, particularly the house wife. Forest grazing is the rule. Tree leaves are extensively lopped for cattle feeding during winter and summer, independently or mixed with agricultural residues or hay conserved on tree tops from the forest or community lands. Only buffaloes are stallfed while dry cows and other animals are left for grazing on hill slopes. Cultivation of forages such as Berseem, Leucerne, Oats, etc. is seldom practised. Though a perfect health cover is essential for higher production, these facilities are seldom available in remote areas. The animals have long generation interval with high age at maturity. In addition to grazing by sedentarised population, these regions support heavy migratory livestock population of *Gujjars, Jads, Bhotias* and *Marchayas.* The forest vegetation, in addition to grazing, is disturbed for temporary hutments made by these nomads.

4. *Proper grazing system:* Deferred-cum-rotational grazing can not be introduced and successfully carried on even for a year. The leguminous component in the existing grazing systems is confined only to low quality annuals. People do not take to stall-feeding except for heavy milch animals like buffalo.

5. *Poor site conditions:* Steep slopes, shallow soil depth, bouldry and gravelly nature of the land, intense soil erosion and low fertility status of the soils in addition to their acidic nature, pose additional problems of development. Indiscriminate cutting of trees and their lopping destroy the normal protective and productive forests of the nearby region. Therefore, these lands require sylvi-pastoral management for fuel and fodder species. The fires, cause dessication of soils, kill the microflora and hamper progressive ecological succession.

Scope for the development of community lands for fodder production is tremendous to improve the economy of the hill region. Besides fuel and fodder, these can sustain several plant based small-scale industries such as fibre and paper pulp, sericulture, soft wood, resins, turpentines, drugs and aromatic oils, etc. Following steps are suggested for increased production and exploitation of the potential.

1. *Enclosure including Soil and Water Conservation Measures:* Stone wall fencing has proved to be most economical and effective for enclosing grasslands from overgrazing in some parts of Western Himalaya though 4 strand barbed wire with angle iron posts have proved most economical in arid zone. Experience has shown that stone wall fencing is economical if the lead for stone is within a range of 30 m otherwise angle iron posts with barbed wire is most economical since the material could be recycled and need very little maintenance cost. Live hedges with *Agave americana* and *Euphorbia caducifolia* could also be tried. Experiments conducted at Bhatta indicate that the above ground production of grasses increased from 21 q/ha to 87 q/ha simply by enclosure for 2-3 years.

Contour trenching at 8m interval with vegetative bunds of hybrid napier recorded a maximum yield of 119 quintals/ha of natural grass and 42 quintals/ha of napier. The vegetative cover showed marked change under closure. *Heteropogon contortus* was gradually replaced by *Arundinella* cover during the period of 5 years. Thus there was a progressive succession in the range.

2. *Brush control:* Closure for grazing initiates progressive ecological succession and thus rapid growth of brush and weeds. The common shrubs recorded are *Lantana camara, Euphorbia royleana, Berberis asiatica, Rhus cotinus, Carissa spinarum* which compete for soil moisture and nutrients. Area occupied by *Lantana camara* only is over 30 per cent in some cases. The number of bushes varies from 21,000 per ha in protected grasslands to 12,500 in unprotected lands. It has been shown that grass production is higher in brush free grassland against grassland with brush indicating adverse influences on grass production. Broadcast of Picloram @ 15 kg/ha around bushes or 0.4 per cent stump application of 2-4 5T ester or amine has proved effective to control bushes. Our studies have shown that spray of Tordon K21 on cut stumps with coarse nozzle completely removes *Lantana* bushes.

3. *Fertilization and manuring :* Studies conducted at Indian Grassland & Fodder Research Institute, Jhansi and Central Arid Zone Research Institute, Jodhpur have shown increase in forage production with the addition of fertilizers, similar results are recorded for the sub-montane regions also. Soil Survey of hilly grasslands have shown nitrogen deficient soils while phosphate and potash very with soil and rainfall. At places due to high rainfall soil becomes acidic and P_2O_5 gets locked in unavailable form. Application of

phosphorus is recorded as less beneficial as compared to nitrogen. Erasmus recorded that 20 kg of nitrogen per ha gave maximum yields over control in *Chrysopogon-Heteropogon* grasslands with a net profit of Rs.92 per ha. For acidic soils addition of 3-5 m tonnes of lime per ha is necessary. Higher doses of nitrogen may be applied only keeping possibilities of ultimate benefits. Studies in Himachal Pradesh have shown that with 100 kg Nitrogen/ha two cuttings gave significantly higher yields.

4. *Choice of species and method of seeding hill grasslands*: Since sylvi-pastoral management for such lands gives the maximum benefit mixtures of grasses legume herbage and trees are recommended. Success will depend upon several factors such as choice of species, soil working, time of sowing and seeding, seed rate and method of sowing. The list of species for the Siwalik region and the outer hill ranges is as below:

Siwalik hill range upto 1,500 m

Grasses:Bothriochloa pertusa, B.intermedia, Cynodon dactylon, Cenchrus ciliaris, Chrysopogon fulvus, Eulaliopsis binata, Heteropogon contortus.

Legumes: Atylosia scarabaeoides, Alysicarpus vaginalis, Clitoria ternatea, Phaseolus trilobus, Stylosanthes humilis, Medicago denticulata, Dolichos axillaris.

Trees:Dalbregia sissoo, Acacia, catechu, A. arabica, Albizzia lebbek, Anogeissus latifolia, Bauhinia purpurea, Bombax ceiba, Bridelia retusa, Cassia fistula, Cordia obliqua, Ehretia laevis, Emblica officinalis, Ficus sp., Garuga pinnata, Grewia oppositifolia, Holoptelea integrifolia, Litsaea polyantha, Mallotus phillipensis, Melia azadirach, Milletia auriculata, Morus alba, Ougenia oojenensis, Putranjiva roxburghii, Terminalia tomentosa, Trema orientalis, Zizyphus Xylopyra.

Outer hill ranges above 1500 m

Grasses: Arundinella nepalensis, Bromus intermis, Chrysopogon fulvus, Dichanthium annulatum, Themeda quadrivalvis.

Legumes: Desmodium uncinatum, Glycine javanica, Lespedeza stenocarpa, Lotus corniculatus, Medicage sativa, Trifolum pratense, T.repens, Trigonella sp. and Vicia villosa.

Trees: Acer villosum, Bauhinia retusa, Cedrela serrata, Celtis caucasica, Cornus macrophylla, Euonymus hamiltonianus, Ficus roxburghii, Grewic opposiiifilia, Morus serrata, Olea cuspidata, Popaius ciliata, Pyrus pashia, Quercus incana, O.dilztata, Robinia pseudoacacia, spermadictyon suaveolens, Ulmus wallichiana.

Singh and Katoch studied the yield potential of a number of grasses from Himachal Pradesh and recorded that the grasses forming the major plant community of natural pastures are relatively low yielders as compared to the exotics. *Setaria Sphacelata* gave 91 percent higher yield than *Pennisetum orientale,* the highest yielder among the local grasses. A number of exotic grasses are under trial at Dehradun. Artificial seeding, reseeding and plantation through rooted slips and tussocks are often necessary for quick regeneration of superior grasses; with only exclosure the process of recoupment is slow.

For effective seeding light soil working is important along the contours though conventional method of sowing is limited to broadcast or line sowing. For the montano region line sowing on steep slopes is not easily accomplished; broadcasting grass seeds with light soil working with a rake ensures successful germination. 1-2 cm seeding depth has given optimum germination. Seed rate is 3-4 kg/ha. Use of unhusked seeds give higher percentage germination over husked seeds. Though premonsoon sowing gives assured establishment of grasses, strong competition from surrounding herbage during monsoon adversely effects forage production. Dry sowing of course has an added significance regarding time availability for such operations.

Planting of rooted slips/tussocks give high and early production, thus compensating for the money and time spent in preparation of slips in nursery. As grass seeds are light and small they need be mixed with moist sand and $CuSO_4$ so that they are not blown away with the wind of eaten away by insects, white ants and rodents. Grass seeds when sown in nursery beds during March-April, will give rooted slips for planting in July. Planting of rooted slips at 1x1 m distance with first effective rains ensures quick establishment. After spot planting, a saucer of 20 cm diameter is made around the plants. From soil conservation view point also this method has an added advantage since the soil is not disturbed over a vast area.

Legumes on community lands are relatively low and of poor quality. It has been estimated that protein content of grasslands could be raised from 2% in pure grass cover to 6% in herbage mixed with legumes. Among the various exotic legumes introduction of *Stylosanthes humilis, Glycine javanica, Dolichos axillaris, Centrosema pubescens and Desmodium* sp. have shown good prospects. Use of clovers, *Medicago and Melilotus* has proved successful in some parts of the submontane regions. Broadcast sowing of pelleted scarified legume seeds before monsoon have been recorded to have excellent establishment.

The technique of soil working, broadcast sowing, spot sowing for legumes, are similar to those given for grasses. In order to have successful growth establishment, legume seedlings raised in polythene bags were found to be effective in this region since competition surrounding herbage is minimised. Grasses and legumes are raised side by side in alternate strips or spots in such a way that 25 per cent of the area is covered with legumes in mixture with grasses. Since the strata of the rock dip vertically and give rise to frequent changes in soil types, from shallow stony to deep clay soil, mixed pastures are successful.

Utilization of field bunds in the Siwalik and terrace risers in outer hill ranges for forage- Singh and Srivastava recorded planting of *Eulaliopsis binata* alongwith *Eucalyptus hybrid* on field bund in the Siwaliks as most economic. A total income from and *Eucalyptus* on bund is estimated at Rs. 1,350/-per ha. important grass used is *Saccharum munja.*

Potential sites which could lbe used for forage production are; (i) Grassed waterways, (ii) river and stream banks and (iii)landslide and cut road faces where only cutting of grasses could be allowed after establishment. Dalal et al recorded under conditions of Dehradun. *Panicum repens* to be the best for grass waterways. However, mixture with legumes like *stylosanthes gracilis* and *Calpogonum orthocarpum* would improve the forage quality. Kudzu provides a good cover on a variety of habitats which can be used as forage when cut. Average yields are 18725, 17725, 13150, 11200 and 4900 kg/ha on pond spillway, nursery, class III, land, roadside slips and torrent banks respectively in the region of Dehradun.

Scarcity of fodder during the rainy season is hardly felt though the quality of forage is poor. Five cattle units per ha has been estimated to be sufficient under ordinary circumstances otherwise eight cattle units can safely be grazed. Peculiar socio-economic conditions prevailing in the region make it difficult to control or reduce cattle unit grazing.

Under the continuous grazing system, while the grass is growing actively, cattle prefer it to legume, promoting new palatable grass growth and allowing the legume to build up a bulk of feed for later use. This process is assisted by the fact that some of the legumes are relatively unpalatable when young. Rotational grazing of grasses at 2-4 weeks interval is helpful.

Yield and persistence of tropical legumes is improved if grazed at 8-12 weeks interval. If the grazing of a grass-legume mixed rangeland is delayed for 8 weeks or more, the grasses will be past its best and cattle will tend to ignore it and concentrate on legume. Under these conditions, grass content is fully utilized while on the other hand shorter durations tend to weaken legume and may essentially eliminate them. The legumes on the other hand retain their protein level for a longer period even after the crop is standing in the field for 6 months.

A protein level of 12-13 per cent may still be expected. Digestibility and voluntary intake of mature legume is much greater than in mature grass. If a range contains half grass and half legume during autumn, it will provide feed of best quality readily acceptable to cattle until new growth comes in March-April.

Table

Crude protein content of some grasses of the submontane region at different tiems of cutting

Grass %of Crude	Protein	during	
May/June	July/Aug.	Sept./Oct.	
Chrysopogon fulvus	8.95	9.60	5.10
Heteropogon contortus	7.44	7.44	4.44
Rothriochloa intermedia	8.12	9.12	8.91

Deferred rotational system of controlled grazing in preferred to continuous controlled grazing. Rotational system entails elaborate demarcation of paddocks and a very careful management which at present in the region is still not possible. However, some elementary form of this system may be practised as being done in some parts of Himachal and Garhwal. These lands are managed as *Ghasnis* under 3 practices viz, (i) enclosure from grazing for 12 months and use it as hay land, (ii) exclosure for six months from June to November, cutting the grass and then grazing from December to May, and (iii) grazing all the year round.

However, the grazing is not carried on the basis of carrying capacity i.e.not controlled grazing. Even then the practices nos. (i) and (ii) have been found to be better than continuous grazing. The air dry forage yield recorded under all the three practices is 3.32, 2.89 and 0.16 mt/year/ha respectively. The Himalayas from east to west present a rich blend of different cultures. The central Himalayam tract between river Kali in the east and river Tons in the west with its innumerable sacred streams and rivers, temples and sacred complexes, legendric peaks and mythological mountains is known as the valley of Gods. Administratively this area comprises of eight hill districts of Uttar Pradesh, referred as Uttarkhand, is divided into two commissioners of Garhwal and Kumaon.

Historically and theologically the people of Garhwal and Kumaon have common roots. For a very long time the entire area was ruled by Katyuri kings. But after fourteenth century the area came under the rule of two different dynasties and as separate states underwent an era of internal strife, feuds and warring kings. Though the area had at times a common king and statesman or huge chunks of migrating populations the strategic ecological, religious, political, economic and geographical factors shaped their cultures into different dimensions.

Sanwal's conclusion that historically and culturally Kumaon and Garhwal have stood as opposed categories and continue to do so to some extent even to-day is true in the sense that the conflicting cultural trends originated in the fourteenth century and different social milieu developed in Garhwal and Kumaon under the Chand and Panwar rulers. In Garhwal strategy and ecology regulated commensal regulations and the inter dining and later connubial exclusiveness played a dominant role in stratifying the social hierarchy. Whereas political power, occupation and economy was considered more significant in Kumaon.

An Anthropo-Historic inquest is made in the following paragraph to highlight the role of strategy and ecology in modulating the inter-dining relations and dietary habits of the people of Garhwal specially the Brahmans And how this commensal exclusiveness later supplemented by connubial interdictions was responsible for stratification of social hierarchy of the Garhwali Brahmans during the last four centuries. The last part of the paper deals with the parameters regulating social mobility among Kumaoni and Garhwali Brahmans.

Garhwal culturally known as Kedarkhand derives its name from the land of forts. After the downfall of Katyuri kings, the area drained by Alaknanda, Bhagirathi and their immediate tributaries was divided into feudal chiefs. The Raja of Chandpur Garhi conquered fiftytwo glens, liquidated the feudal chiefs and after consolidating the entire area under his rule shifted his capital to Srinagar in 1517 A.D. He was the thirty-seventh ruler of the Panwar dynasty but the first king to rule over united Garhwal from a central capital.

In order to control a large territory and remain prepared against uprisings by the vanquished glen chiefs he deemed necessary to have a standing army at

Srinagar. Commensalism posed the first obstacle in maintaining a regular army. In those days the upper class Brahmans and Rajputs constituting the Garhwali army observed the rules of inter-dining exclusiveness. To solve the problem of individual cooking, Raja Ajaipal appointed twelve Brahman kin-groups as official cooks and issued an enjoinder to all Brahmans formed by these twelve Agnatic kin-groups became acceptable cooks in social and religious gatherings also and became famous as Sarolas.

The twelve patriclans of the Sarola class had accompanied the royal entourage from Chandpur Garhi, were already in the service of royal household and thus confident and trusted people. Their choice was obvious. Raja Ajaipal could not afford to risk the appointment of newly recruited Brahmans from the Srinagar area. The new cooks could prove unreliable and pose security risks. They could fall prey to subversive elements, poison the army or create conditions for mutiny and desertion.

Naturally the choice or military and royal cooks was limited to the trust-worthy and old acquaintances till the new subjects proved their loyalty in the course of time. For over hundred years only twelve kin-groups constituted the Sarola class. In 1629 A.D. when the Garhwali army was stationed at Daba in Tibet to take retaliatory action against the invading Tibetan forces, a necessity of more cooks was felt and Raha Mahipat shah added six new Brahman kin-groups to the Sarola class taking their number to eighteen. Later three more patri class were added to their class.

The war requirements were responsible for the increase in number of Sarola kin-groups. From 1500 A.D. when Raja Ajaipal ascended his throne to Raja Pradyumn Shah who was killed while fighting the invading Gorkha army in 1804, the Garhwalis fought eighty-five battles within a span of three hundred years. Suiting the needs of mountain warfare and the rules of commansality thirty-six agnatic kin-groups composed the Sarola class of Brahmans by the end of nineteenth century. The first batch of twelve agnatic kin-groups appointed to the Sarola class belonged to the villages surrounding Chandpur Garhi. But the latter inductions or appointments were instantaneous and based on purely strategic reasons.

When Raja Mahipat Shah raised their number from twelve to eighteen at Daba the kin-groups whose male members were locally available on the spot at the battlefield were included into the Sarola class. The initial appointment and subsequent increase in the number of Sarola groups illustrates that the Sarola class came into existence on piecemeal and makeshift arrangements at different intervals of history and was not a complete conversion of one class of Brahmans into this classification. All patriclans named as Sarola initially or subsequently, existed as members of some other groups of Brahmans before joining the Sarola group.

The initial twelve clans were residents of villages surrounding Chandpur Garhi but the subsequent clans were from Srinagar area and mainly belonged to Gangari group. The Sarola Brahmans as a class had no reservations on food cooked by the fellow Sarolas but laid interdictions on Kachi Rasoj cooked by other Brahmans. Compared to Sarolas, the Gangari class of Brahmans is composed of prominent, high class and more number of Agnatic kin-groups. Both the Sarolas and Gangaris claim themselves immigrants to the hills. In fact some of the kin-groups belonged to Gangari Brahmans, held major and responsible posts in the Royal Court, Army, social and ecclesiastical circles but did not eat Kachi Rasoi cooked by fellow Gangaris. They engaged Sarolas to cook in their houses in marriages and other social, religious and suspicious occasions so that all assembled Gangari, kin-groups could eat together. The Sarolas took advantage of the commensal privileges and by the end of eighteenth century established their superiority over the Gangaris. It is beyond the scope this paper to discuss in details the etymological roots of the word Sarola. Whether it is Shrotriya, Seera, Shrestha or even Raswala the feeling of commensal superiority and distinct identity conceived the idea of purity of class among Sarola Agnatic kin-groups.

The Sarolas imposed connubial exclusiveness to maintain their purity and class distinction and became and endogamous class. Marriages outside the class wedlocks lost class status. The marriages with Gangari spouses were permitted but they were considered some sort of morganatic or non-sacramental marriages which took away the crucial privilege of paternal group affiliation and the off-springs belonged to a new Gangari-Sarola class of Brahmans.

The socio-cultural milieu in Uttarkhand is deeply shaped by Himalayan ecology. The mountainous terrain and the cold climate not only regulated the dietary habits but also enforced certain dress and clothing requirements. The ecclesiastical canons of the Indo-Gangetic Brahmanical order did find favour with the Garhwali Brahmans for sometime but in the course of history the mountain ecology modulated the prohibitions on cooking, serving and eating of Kachi Rasoi and the dining dress regulations. The Garhwalis are basically rice eaters but the preparation of rice was confined to day time only.

The cold and dark evenings discouraged the cumber some ritual of stripping upper garments before eating rice. Moreover on social and religious occasions the rice was cooked in open fields or courtyards in bright sun and not in the house kitchens. In evenings the rice was substituted by chappaties prepared from wheat flour, maize or Rye corn. The chappaties are considered pure and do not form a part of Kachi Rasoi since last few centuries. This separation of chappaties an common vegetable from Kachi Rasoi in Garhwal was due to ecological and strategic necessities. Till the beginning of early seventeenth century Kachi Rasoi in Garhwal included both Bhat and Roti. During those days the Kachi rasoi was eaten with three specific reservations. It must be prepared by a respectable and acceptable cook. It was to be consumed within the specific sanctified periphery of the kitchen or the place where the food was prepared. And except for the loincloth all other clothes, specially the upper garments must be stripped off.

In 1629 a Garrison of Garhwali army was positioned at Daba to take punitive action against the Tibetan invadors. In spite of severe cold the Garhwali soldiers were stripping upper garments at the meals timings. The commander of the Garhwali army foresaw the dangers of possible sickness in that cold and advised the king to modify the rules of eating Kachi Rasoi. Raja Mahipar Shah in consultation with his Dharmadhikari issued a royal decree making Roti Shuchi. The royal order meant that henceforth the Roti and common vegetables cooked by respectable Brahmans and Rajputs can be eaten outside the periphery of the kitchen and without removing the upper garments.

Thus after 1629 the Kachi Rasoi was divided into two parts viz. the Bhat and the Roti. Due to royal decree and strategic and ecological necessity, the practice of eating chappaties without removing upper garments was slowly adopted all over Garhwal. The Chappati also became a convenient menu for the evenings. The interdictions on rice remained intact and infact the prohibitions imposed on rice eating was the main differentiating factor between the Sarola and Gangari class of Brahmans. Ecology has another surprising impact on hill Brahmans. While they refused rice from each others hands, following the customs of Trans-Himalayan tribalism, they have no inhibitions in accepting mutton or other acceptable meats ad intoxicants from members of other or lower class of Brahmans or Rajputs. Even a high class Sarola not accepting rice cooked by his Gangari Brahman wife could eat mutton prepared by a Rajput or a lower class Brahman.

Comparatively the Nepalese confronted a similar situation during the time of Raja Prithwi Narayan Shah and liberalised rice eating at the time of war. During the siege of Kathmandu city, the Raja found that, stripping of upper garments may come while the army is consuming food. So he ordered that in the battlefield even cooked rice could be eaten without removing the clothes and till today the Nepalese soldiers followed this practice. The Garhwali army never confronted such situation or received any royal decree to relieve them from the pre-requisites of stripping garments before eating rice.

The difficulties created by commensal exclusiveness and the dress regulations among the Garhwali Brahman soldiers were also experienced by the British Commanders and as a result they stopped their recruitment in Garhwal rifles. The recruitment of Brahmans was stopped in Garhwal Rifles in 1872 and opened only on the eve of independence in 1947. However during the British days a number of Gangari Brahmans, who has no reservations on rice eating served in Garhwal Rifles by enrolling themselves as Rajputs.

The story of reducing the original inhabitants of Kumaon and Garhwal to a state of slavery by Khasi has been accepted by various historians and sociologists. The classifications between clean and unclean people in

Binary discriminations holds good in both Kumaon and Garhwal as the doms both artisans and agricultural labourers were landless people and were regarded as polluting. The migrant Syndrome as status symbol divided the Brahman and Rajput society.

The immigrants considered themselves higher than the indigeneous people. The social structure of Uttarkhand which saw its apex in the end of eighteenth century did not change much with the advent of British rules and infact its remnants are still active in the rural interiors of Uttarkhand even today. The factors responsible for social stratification in rural Kumaon as brought out by Sanwal hold good in Garhwal also, except the upper class Brahman society. The Brahminical societies in Kumaon and Garhwal differed drastically in their mobility and status symbols.

The Brahmans in Kumaon were not only the political and economic policy makers but the defacto rulers. After the Nepalese expulsion the British administrators did not interfere with the Mumaoni social structure. On the other hand, Garhwal though bifurcated after 1815 was ruled by a king till 1950. Moreover, the Brahmans in Garhwal were mainly sacerdotalists at the favour major Tirthas, their allied shrines and numerous sacred complexes. The operation of commansality in Kumaon was an important mechanism for maintaining group exclusiveness but as a symbol of status equality it was not important whereas in Garhwal as a consequence of strategic and ecological factors, the hierarchy born out of commensal exclusiveness always maintained its superiority.

The Sarolas after proclaiming themselves a distinct class did not open their avenues for influential Gangari Brahmans holding powerful political and economic positions in late eighteenth or nineteenth century. None of the caste groups in Kumaon except the Doms were completely closed groups. The Pachbiri Brahmans who were at the middle of the hierarchy in Kumaon acted as reserve for Chauthani brides. Initially the lower Brahmans and the social mobility in Kumaon was always flexible. But in Garhwal it was a rigid affair and the entry into Sarola class was only by birth. The Sarola class was formed by the royal decrees and some of their patri clans initially belonged to Gangri class, but once they became a distinct class, except for one or two rare cases of adoption or marriage, there was no entry by virtue of political and economic gains or regular hyperganic marriages.

The Sarolas tried to make their class as endogamous and matrimonial alliances with Gangri spouses were considered as non sacramental marriages. Thugh they did not succeed in this enfeavour. The shortage of connubial alliances among the permissible patri clans forces them to seek Gagari spouses. These marriage were considered as morganitic. The children born out of such wedlocks lost the privilege of paternal group affiliation and gave rise to a new class of Sarola-Gagari Brahmans. The Sarola patri clans were divided into classes of pure Sarolas and Gagari Sarolas.

The latter group except inherting the Agnatic Kingroup titles of Sarola patriclans were part and parcel of the Gagari class of Brahmans. There was no discrimination in occupational or bureaucreatic appointments among Sarolas and Gangaris. Statistically the Gangaris held higher and more influential posts in royal courts. Eight of the Navratnas in the court of Maharaja Fateh Shah belonged to Gangari group.

A large number of Faujdars, Dewans and even the Dharmadhikaris were Gangaris. But a Sarola could have a Gangari guru, receive/ordain the Yagyopavit, marry his daughter but could not beget a Sapind child from the

Gangari spouse. Connubial alliance with a Gangari girl was a hypogamic marriage for a Sarola and resulted in loss of status for the children, whereas it was not a hypergamons alliance for a Gangari to marry a Sarola girl and the children maintained status quo.

A group of four patriclans of the Gangari class attempted to make distinct class of Brahmans as chothokee class, as members of these four agnatic kingroups held influential posts in the royal court. In Kumaon one could enter the topmost chauthani class after attaining an influential post in not by marriage. But the chothokees in spite of their hereditary bureaucratic positions remained a part and parcel of Gangari class due to commensal and connubial limitations. Comparatively the Gangaris were an open class and added new patriclans into its hierarchy at various times.

Unlike Kumaon there were no fixed norms for upward or downward mobility between Gangari and other lower class of Brahmans. The Gangaris at times married their daughters to economically well to do or educated grooms belonging to Khasi Brahman patriclans and by way of these hypergamic marriage added to the number of Gangari kingroups. At times the Gangari Brahamans also married Khasi girls but in most cases did not accept rice cooked by wife's kinsmen. Once a member of the higher echeleon like the Gangari in Garhwal or pachbiri or Pitali in Kumaon, there was no downward mobility to Khasi class even on political, economic or educational status loss.

The Khasi class of Brahamans and Rajputs do not exist today. By middle of this century all Khasi Brahman patriclans had become a part of the Gangari class. Patriclans from the Vaishnavite and Shivite mendicants. Nirola class of Brahamans, Deo Prayagi an Kedar Pandas and Devotee class of Brahamans are making inroads into the Gangari class. In spite of their priestly and sacerdotal occupations the Sarolas enjoyed higher status in the parochial society. Whereas the secular Brahman enjoyed a higher status than his priestly compeer and the pujaris and purohits have every where in India a lower status than Brahamans, who hold lucrative positions.

The chauthanis in Kumaon considered themselves superior to pachbiris on this assumption. But in Garhwal the attempt made by Chothokee class of Gangari Brahamans holding hereditary bureaucratic and influential positions failed to make any impact on this higher to the Sarola class? A number of factors are responsible for this. They formed a composition and well knit class in which the weaker kingroups or patriclans were helped by their influential fellow kingroups. Their social unity was well organised and kept a track of its kingroups. In 1910 the Sarola Sabha notified that some patriclans ceased to be pure Sarolas as they do not have a single family of pure Sarola class. Some of their patriclans held hereditary posts of Rajgurus, Naibrawals and other ecclesiastical orders.

Some of them held influential and powerful positions in the royal court and combined in them the religious and bureaucratic superiority. Lastly with so many sacred temples and rivers, pilgrimage being the major source of economy, sacerdotalism being the main occupation of Brahmans in Alakanda and Bhagirathi valleys, the natives could not afford to lower the hierarchy of the priests of the valley of Gods.

The present communication reports different ecological problems in the Garhwal Himalayas. Attempts have been made to review the pertaining literature and it is concluded that there is an urgent need to develop such studies so that different problems of the region could be solved. From such a point of view various habitats and ecological factors have been recognized and discussed. In order to check further deterioration of the climatic conditions and ecology, a number of suggestions have been put forward. A review of the pertinent literature reveals that this glorious region of the Himalayan terrain has received considerable botanical attention especially for floristic survey. However, still there are several regions which demand detailed plant explorations. For instance, very limited, if at all, attention has been paid to investigate the cryptogams, lower vascular plants, and the aquatic flora of the area.

Osmaston was perhaps the first person to emphasize the need for detailed ecological study of the Garhwal Himalayas and his plea has been repeatedly supported by Dudgeon and R.K. Gupta, whose own studies were restricted to the description of forest communities only. More recently, Saxena and Srivastava have published the results of their analysis of the forest communities of Mussoorie in relation to the topography and climate. Nevertheless it can be safely concluded that the ecological studies of different biomes at variable elevations of Garhwal Himalayas have received little attention chiefly because of relative inaccessibility. If we move the mirror towards the historical background of ecological studies in different parts of the world, we find that during the first three decades of the current century, there started to develop such specialised branches of ecology as palaeoecology, zoogeography, oceanology, limnology, etc.

Simultaneously, population concepts were introduced for which statistical studies and the community problems. After the introduction of the term 'ecosystem' in the literature by Tansley, quantitative attempts were stimulated which could lead to the development of the concepts of productivity and energy relations. Finally, the present status occupied by ecology is quite evident by its role in the solution of socio-economic and political problems such as those of soil conservation, pollution, field-management, and space. A brief pause to contemplate the aforementioned facts and a review of literature on ecology of Garhwal Himalayas clearly indicates the need to develop such analyses with their fundamental worth so that we are able to solve various problems of this region to aid in the current trends of advancement in different spheres of national endeavour.

Geographically, the Himalayas have been classified into four divisions: 1. Eastern or Assam Himalaya. 2. The Central or Nepal Himalaya. 3. The Kumaon or Western Himalaya. 4. The North-west or Panjab Himalaya. The Western Himalaya extends for about 320 kms between the river Kali and the great basin of Satluj. All the eight districts of Kumaon and Garhwal lie within this region. As such this Himalayan part can also be termed as Uttarakhand Himalayas and Garhwal Himalaya is a part of this division, restricted to five districts of Chamoli, Dehra Dun, Pauri, Uttarkashi, and Tehri.

Up to an altitude of 1,000 m the zone is characterized by the dense forests of *Shorea robusta*; mixed deciduous forests with *Anogeissus latifolia, Dalbergia sissoo, Lagerstroemia parviflora,* and *Terminalia* species; swamp forests with *Bischofia javanica,* Carallia brachiata, Salix tetraserma, and Pyrus pashia as the chief components, with occasional cane breaks. The moist slopes are occupied by species like Cedrella toona, Eugenia jambolana *and* Ficus glomerata.

The isolated patches of grassland vegetation chiefly consist of Eulaliopsis binata, Heteropogon contortus, Botheriochloa *sps.,* Dichanthium *sps.,* Eragrostis *sps* Eragrostis sps. and Chrysopogon sps. Above 1,000-1.200 m, usually on dry locales, Pinus roxburghii, forms excellent forests reaching up to elevations of more than 2,000 m. The drier belt, however, shows, xeric elements like Acacia *spp.,* Mallotus philippinensis *and* Zizyphus jujuba, with dense patches of thorny succulents such as Euphorbia royleana on highly steep slopes. Here the grass-land patches are also frequently scattered on saddles and ridges where Arthraxon prinonodes, Arundinella setosa, Erianthus, rufipilus, Eulalia quadrinervis, Imperata cylindrica, Sporobolus diander *and* Themeda anathera, are the predominant grasses.

In between the altitude ranges of 1,200 and 1,500 m pine is allociated with Quercus incana *and* Rhododendron *species.* Above 1,800 m, the northern slopes with moist situations are covered with oak, Rhododendron forests interspersed with Myrica farquhariana and Lyonia ovalifolia. The vegetation at the higher altitudes exhibits a dense canopy of Quercus floribunda at moist situations, which occupies an intermediate range between Q. incana and Q. semicarpifolia. These forests are also rich in epiphytes, orchids and ferns with lianaceous climbers especially Hoeboellia latifolia, Schizandra grandiflora and Vitis spp. are common. Above 2,800m, Oak-Conifer association occurs where Quercus semicarpifolia, Abies pindrow, Rhododendron arboreum, Taxus walllichiana, Pyrus and Viburnum species are the dominant forms. One can also witness here the inner drier region in the altitude range 2,000-3,00 m with forests of Cedrus deodara. The outer range, however, exhibit pure forests of Curpressus torulosa.

Nbetual utilis and Abies spectabilis occur either singly or mixed in various proportions extending up to 4,000 m. In several localities Quercus semicarpifolia-Abies pindrow forests lying south of the snows as well as the Cedrus deodara forests north of them, merge into the above type of forests. Betula is associated with Rhododendron campanulatum at higher altitudes. In subalpine forests of Betula-Rhododendron as well as the shrubby areas of Rhododendron anthopogon, R. lepidotum, species of Cotoneaster, Lonicera, and Rosa are better developed on the north exposed slopes. In moist and open localities, luxuriant meadows of herbaceous perennials are seen with the usual components represented by general such as, Anemone, Epilobium Gentiana, Geranium and Polygonum etc. The important meadow formations are Auli and Gorson near Jamnotri, Bendi, and Kush-Kalyani etc. The curious, woody species of Saussurea are widely distributed on the moranic slopes.

At several localities in Garhwal, extremely dry conditions prevail and here the prevalent plants belong to species of Berberis, Caragona, Ephedra, and Juniperus. The richest herbaceous alpine flora is exhibited by Bhyundhar Valley of North Garhwal. The details about vegetation patterns of higher peaks of Garhwal are still unknown, nevertheless records from regions of similar elevations have been described by Swan who found fungi and yeast in soil at elevations of 7,600-8,400 m. Ecologically, this zone may act as an important area since it regulates the meteorology and hydrology of all the surrounding regions.

Apart form terrestrial habitat, the Garhwal Himalayas comprise a number of lakes such as, Deoria Tal, Sahastra, Bhil Sarovar, Vasuki Tal, Gandhi Sarovar, Satopanth Lake, Hemkund, Gaurikund, Roopkund, Tammakund, etc., roughly between the elevational ranges of 1,650 to 5,000 m. The important rivers like Alaknanda, Dhauli, Mandakani, Pindar, Nandakani, Bhilangana, Bhagirathi, Eastern and Western Nayar are additional systems where vegetation needs detailed explorations. Major biomes of Garhwal Himalayas can be classified, on the basis of the kinds of habitat, plant association and environment conditions, into the following types:

Aquatic habitat

This consists of a number of lakes at different altitudes and rivers. Plant communities occurring in this habitat can be divided into two systems:

Lentic ecosystem

Includes the habitat with standing water as different kinds, tals, and sarovars in this area, and

Lotic ecosystem

Comprising running water habitats scuh as as rivers, etc.

Terrestrial habitat

This part includes the range of mountains encompassing most important botanical regions with climate and vegetation ranging form truly montane to temperate to alpine type from low the high altitudes in successive belts. The important biomes associated with this habitat are:

The high altitude biomes

Including the regions of semi-attenuated atmosphere, low atmospheric temperature, high aridity and harsh contrast between macro and micro-climatic conditions. As such, the type includes the whole of the elevated region above the timber line on Garhwal mountains. This biome consists of: (i) Dwarf bushy ecosystem, (ii) Herbland ecosystem, and (iii) Meadowland ecosystem

The sub-high altitude biomes

These cover the regions between the timber-line and the basal zone of conifer commencement and characterized by forests, grassland patches, as well as shrub and herb communities which can be divided into following systems:

(i) The moist woodland, (ii) The dry woodland, (iii) The grassland, (iv) The shrubland, and (v) The herbland.

The low altitude biome

This covers the regions below conifer belt including foot hills, Tarai and Bhabar and includes: (i) The moist deciduous woodland ecosystem, (ii) The swampland ecosystem, (iii) The mixed deciduous woodland ecosystem, (v) The grassland ecosystem, (c) The shrubland ecosystem, and (vi) The herbland ecosystem.

Ecological factors

Various factors influence the organisms in a holiastic manner, i.e., these operate in conjunction and not in isolation to each other. The different altitudinal zones of Garhwal Himalayas are chiefly affected by two categories of ecological factors:

The abiotic factors

Mani reported that Himalaya profoundly influences the climate and vegetation of the surrounding lowland region. It must be simultaneously suggested that the prevailing climate and the vegetational characters of lowland more or less greatly modify the altitude-conditioned environment of the mountain system.

Truly speaking, the mountain is composed or ridges and saddles, and two opposite slopes meet either with river, rivulet or lake, Each slope of a particular altitude has its specific topography, influenced by a number of factors and the physico-chemical nature of soil is largely dependent on the parent rocks composing it. The submontane, temperate and alpine zones show irregular topography with variable soils along-with microclimatic conditions which control the distribution pattern of vegetation.

In spite of the activities of different microorganisms,

animals and fire, different biomes of Garhwal Himalayas are chiefly influenced by the following activities of the natives: (i) The domesticated animals of the natives like cow, ox, sheep, buffallow, goat, etc., damage most of the vegetation which ultimately change the soil condition. (ii) The excessive felling of different valuable and ecologically important vegetation by the natives and the forest contractors causes marked changes in biomes. Irregular felling of vegetation recently caused invasion of different weeds as *Lantana camara, Oxalis* spp., and *Tagetes* spp., etc. It is well-known that invasion of conifers in Himalayas has taken place due to irregular felling of climax loak forests:

We are making the following suggestions for checking further deterioration of the habitat:

1. Detailed ecological survey of different vegetation types in relation to their co-habitat at different altitudes to establish ecological relationship and gradient.
2. Study of phenological behaviour of different ecosystems to correlate relationships between eco-factors and vegetation in relation to physiognomy.
3. Evaluation of various plant community structures at different altitudes in a comparative manner to find out the adaptability of individuals in the community and niche differentiation; establishment of basic parameters of different ecosystems; their structure and function.
4. Survey of scavengers and secondary producers and their role in different ecosystems.
5. Production behaviour of different communities at different altitudinal zones along with structural modifications in ecological amplitude and relation to microclimate by establishing peak growth biomass and net primary production, The study of biomass

of the flora and Fauna in its entirety and interrelationship.

6. Energy conserving efficiency and budget of different communities at different elevations.
7. Study of different communities influenced by the biotic disturbances and their relationship with undisturbed communities.
8. Determination of the ratio of each sere or segment and terminal standing crop to annual net production.
9. Experimental studies of succession by encompassing analysis of (i) litter (ii) shade (iii) shaded litter, and (iv) reduction in root competition on a regional basis in space and time. Simultaneously Continuum Index be studied for succession on the basis of relative frequency, density and abundance.
10. Observation of competition behaviour of different weeds.
11. Autecological studies of economically useful species in various directions to enable an understanding of the environmental and nutritional requirements of these plants which would help in propagating important food and fodder plants growing in this region.
12. Knowledge of steps of lttler formation and decomposition in different biomes and its importance.
13. Study of physico-chemical nature of different soil types at different elevations and analyses for different elements in soil and plants to study the mineral cycling.
14. Effective pollution measures in different ecosystems.
15. Study of alpine ecosystems in relation to climate, flora and fauna along with its influence to lower altitudes.

16. Study of ecological approaches made by ants and spiders.
17. After evaluating and collecting the results from different biome,s improvements measures can be undertaken, considering the difficulties of the natives.

Despite its strict technical usage to elucidate the process of interaction between nature and living beings, the concept of 'ecology' now-a-days is often used interchangeably with 'environment'.

Of the multiple constraints, peculiar to Himalayan ecology, altitude, gradient, snow, rainfall, wind and temperature and some noteworthy factors and their composite or individual effect create an impact- positive and negative, among the inhabitants. They feel constant insecurity due to dangerous tracks, forests full of lurking beasts, varieties of natural upheavals, climate extremes, crop failure, disease and death etc., different factors of their habıtat. These elements of insecurity at every step have made Himalayans to be God- fearing. They are so scared of their ecological setting that even the sound produced by the breeze on the dry leaves in the lonely forests is enough to accept the presence of some malevolent spirit. In this manner the Himalayans ought to perceive a number of deities and spirits, one behind each of the ecological happenings and pace them in their pantheon.

Like wise liturgy, shrines, religious performances, specialists, and ideology etc., various aspects of religion in Garhwal Himalaya reflect ecological bearing. The present article aims to delineate different elements of ecology responsible for originating faith in deities and liturgy adopted for sacred performances, in Garhwal Himalaya, as evident from the field studies from Badeti, Deora and Dharali villages of Uttarkashi district of Uttar Pradesh.

The district of Uttarkashi is a part of erstwhile Native Kingdom of Garhwal, and belongs to Central Himalaya, having a tangled series of mountain rides with innumerable spurs separated by narrow valleys. The mountain ridges radiate from a loafty series of peaks boarding Tibet at the hight of 20,000 to 23,000 feet above sea level. Snow melt water and rain water is drained by a number of streams and rivulets, which together give rise to three main river systems-the Bhagirathi, the Yamuna and the Tons and three synonymous valleys.

The valleys are deep and narrow and mountain ridges have steep gradients. On the basis of rainfall, elevation, wind and temperature etc., varients, the habitable mountain area of the district is to be categorised into three broad vertically graded climatic zones. The first being the lower altitude or the Monsoon zone with tropical climate; the second is the middle altitude of the Monsoon zone with temperature climate, and the third the high altitude or the Arid zone with colder climate.

During summers everywhere the places exposed to sun become warm at the day time while nights are cold throughout. Snowfall occurs as low as 4,000 ft., in winters. Interplay of local configuration and climate is largely responsible for the variation in seasons in the three altitudes. In middle and lower altitudes there are three seasons- the *Sitkal* the *Rudi* and the *Vaskal*. However, in middle altitude *Sitkal* persists a little longer and Rudi lessens. But high altitudes enjoy only two seasons- the long spell of *Sitkal* of *Heund* and summer or spring in the rest months of the year.

Several climate-ecological constraints of the Himalayan topography outlined in the preceding paragraph incessantly affect the cultural milieu of the people, of which religion is a part. Religious faiths and practices evolved as natural instincts of adjustment against the adverse environmental conditions prevail at every nook and corner of the habitat. Usually, the ecological factors and happenings beyond the controlling capacity of mankind are thought to be controlled by supernatural forces, which are deified and perceived in the forms of deities, manes and other benevolent or malevolent spirits. They are deligently propitiated or appeased by placing them in pantheon and arranging regular offerings to ward off them in pantheon and arranging regular offerings to ward off the dangers associated with their influence on one hand and on the other to secure prosperity, progeny, health ad happiness.

Ecological bases for deification and beliefs, specific to each supernatural force are detailed in the following paragraph. Rocky ridges of Garhwal Himalaya present steep gradients of uneven surface having a few meters of soil crusts on these and that too is highly erosion-prone. All the activities of life including habitations and food producing strategies solely rest on the soil. Realizing the utmost importance of soil and instable surface, people have deified it in the form of *Bhumia* or *Matia* protects soil erosion and land-slides during seismic thrusts and excessive rains etc., natural calamities.

Visiting over *Danda* or *Kantha* for tending cattle, collecting fuel and fodder and other items is a routine works of the Himalayans. But these tops are lonely, insecure and full of a kind of fragrant air, which make the visitors unconscious. To cope with such circumstances, people believe in a supernatural force called *pari*, an offer the offering when people feel the attack of *pari*. People of Garhwal have long association with *Dhangaar* where they live, walk, play and do agriculture, always retain the risk of slipping down hundreds and thousands of feet below the surface where death is inevitable. To avert these unexpected incidences,

Goddess *Matri* or Achhri is worshipped. Multiplicity of *Khala* and *Gadhera* in the Himalayan topography has again created troubles and life risks for the people and cattle, where they usually visit for different purposes. To overcome troubles associated with these phenomena people believe in Jharmand and place in their pantheon.

Places like *Margad* and confluence of rivers are believed to be very fearsome because these are frequented by demons and spirits which harm the people passing nearby. To save themselves from these miscreants, people appease the deity *Bhainrav* and give timely offerings. Garhwal is full of forests and they are the part and parcel of the life of the people. They largely depend on the forests for the supply of fuel, fodder, building material, medicinal herbs, honey, gum, spices, fruits, berries and roots etc., many importance forest produces.

Besides, forests give shelter to a variety of animals and birds which are the constant source of meat and other valuable products to cater local needs. For this inseparable dependence on the forests, people have placed these as deity Banakya in their pantheon and worship for appeasement. Mountain topography with its enormous vegetation and caves provides suitable habits to flourish a great variety of animals and harmful creatures. To get rid of the lethal attacks of these wild animals on themselves and their domesticated cattle, people worship a deity called *Narsingh*.

Abundance of snakes and other poisonous creatures in the Himalaya is always a terror for its people and cattle. To get rid of snake bites and the bites of other poisonous creatures God *Nagraja* is appeased by the offering of milk and milk products. It is believed that the deity also protects cattle, increase their progeny and milk in them. Being the homeland of vegetation the Himalaya also houses varieties of Bacteria and virus which cause different types of diseases in human beings and cattle in which dangerousmost are epidemic because they dispeople the villages.

For the effective control of such diseases Goddess Kalka is propitiated by placing in pantheon and dedicating different offerings. Long Himalayan winters, during which most of the territories remain frozen under heavy crusts of snow and the rest face chilly cold, completely paralysing the life of the people. This ordeal is finished by the warmth of the rays of the sun, so people attribute additional importance to sun and place it as Sun-God in their pantheon and appease through various performances.

Incidences of God and thunder bolt are very frequent in Garhwal Himalaya. They ruin life and crops with in shorter time. To have an effective check over such natural occurrences deity Bijar is worshiped at regular intervals. Himalayan eco-system is very treacherous and creates critical conditions through sudden earth-quakes, floods and draughts. To combat these conditions lord Shiv is worshipped and appeased. Lord Shiv brings rain, stops food and seismic thrusts.

The vast and tough territories of Himalaya are very clam and quit. In utter silence lone individual even in this house feels the presence of some invisible forces. These forces, it is said are malignant and termed as Blunt and are propitiated when someone feels their influence.

Like religious beliefs, liturgy too emanates ecological elements. Liturgy followed to observe rituals in Garhwal is almost similar for all excepting the nature of offerings. The most striking feature is the practice of exorcism and predictions about the happenings of nature with whom Himalayan life is tagged. Playing of different types of games, and days and nights long collective dances of

Raso and Tado regardless of sex and age are other notable features. All these constitute essential part of the rituals, besides, the offering of flower, milk, water, incense, Akshat, Diva, Halwa, Puri etc. arranged by common contributions or efforts on village level as in Badeti and Dharali or on Patti level as in Deora. Playing of competitive strenuous and lithargic games during rituals such as Genda and Hindoda traditional to Deora and Badta todo, Latha todo, Bhela or Ola in all places, lays emphasis and encourages people to keep their bodies fit and energetic to sustain hazardous mountain life. These games also provide people with a source of entertainment in monotonous mountain life.

Likewise ritual dances and songs of different kind, give people enough entertainment and rejoicing, During dances people forget all sorts of worries ecological or otherwise and feel fresh and ready to combat with the adversities of their terrain. The chilly nights of winters, when atmosphere remains frozen, is passed by such dances around fire-places. The month of Poos has a chain of festivities in continuation for nine days. These nine days are passed by enjoying lavish dishes and various enjoyments to lessen the harshness of climate.

The time of the performance of most of the religious rituals mark climatic change, and natural happenings. The rituals conducted on the Sankranti of each month mark the passage of Sun from one constellation to another. Although each Sankranti day in Garhwal is considered to be ritually important still the Bissu the Sankranti of Mekh in the month of Magh have greater importance. Because they respectively mark Dakshinayan and Uttarayan the two important seasonal conditions. Since Uttarayan heralds an increase in day span and consequently warmer days to break ice-cold ordeal of winters, it is welcomed by celebrating festivals for nine days in continuations in the month of Poos. Similarly, festival of lights marks onset of winter season. In the lower and middle altitude the festival of lights is called as Baqwal and celebrated for two days from 14th to 15th of Kartik i.e., Amavas. But the festival of lights in high altitude is known as Shelku and celebrated quite earlier for two days from Bahdon Masant to Asoj Sankranti, in the similar pattern.

Nag Raja is worshipped two times, once on any Wednesday and second time in the month of Sravan on Nag Panchmi day. Both the times mark rainy season, when snakes and other poisonous creatures usually come out of their hides and loitre on the surface. This is the time of their mating also. So, for obvious natural reasons they become furious and to avoid their bites deity Nag Raja is appeased at proper time. Likewise, other deities which are responsible for different ecological constraints are also propitiated before the time of happenings and in very few case after the occurrence of accidents in their topography for the effective control of further incidences.

Attributing special importance to exorcistic performance in Garhwal also reflects ecological grounds. The hazardous terrain, they cover to reach rivers for sacred dip along with their deity palanquine and come back to religious shrines is really very challenging task for the Himalayans. Such strenuous ventures are eased when they are possessed with the charge of their pet spirits. When charge of the deity possesses people, they take palanquine of deity on their shoulder and turn towards destination and they are followed by the villagers who for simple curiosity forget or say overlook the risks and fatigue of their way.

The exorcistic performances taken into account in Garhwal are Pandav Nantch, Dhol, Doli, Jagar, Ghadyala and Bedarth Radana. They are performed to avoid natural calamities, diseases, and unwanted natural risks. These performances provide good entertainment to people, who enjoy these ignoring crucial factors of ecology.

The present communication reports different ecological problems in the Garhwal Himalayas. Attempts have been made to review the pertaining literature and it is concluded that there is an urgent need to develop such studies so that different problems of the region could be solved. From such a point of view various habitats and ecological factors have been recognized and discussed. In order to check further deterioration of the climatic conditions and ecology, a number of suggestions have been put forward. A review of the pertinent literature reveals that this glorious region of the Himalayan terrain has received considerable botanical attention especially for floristic survey.

However, still there are several regions which demand detailed plant explorations. For instance, very limited, if at all, attention has been paid to investigate the cryptogams, lower vascular plants, and the aquatic flora of the area. Osmaston was perhaps the first person to emphasize the need for detailed ecological study of the Garhwal Himalayas and his plea has been repeatedly supported by Dudgeon and R.K. Gupta, whose own studies were restricted to the description of forest communities only. More recently, Saxena and Srivastava have published the results of their analysis of the forest communities of Mussoorie in relation to the topography and climate.

Nevertheless it can be safely concluded that the ecological studies of different biomes at variable elevations of Garhwal Himalayas have received little attention chiefly because of relative inaccessibility. If we move the mirror towards the historical background of ecological studies in different parts of the world, we find that during the first three decades of the current century, there started to develop such specialised branches of ecology as palaeoecology, zoogeography, oceanology, limnology, etc. After the introduction of the term 'ecosystem' in the literature by Tansley, quantitative attempts were stimulated which could lead to the development of the concepts of productivity and energy relations.

Finally, the present status occupied by ecology is quite evident by its role in the solution of socio-economic and political problems such as those of soil conservation, pollution, field-management, and space. A brief pause to contemplate the aforementioned facts and a review of literature on ecology of Garhwal Himalayas clearly indicates the need to develop such analyses with their fundamental worth so that we are able to solve various problems of this region to aid in the current trends of advancement in different spheres of national endeavour. Geographically, the Himalayas have been classified into four divisions:

1. Eastern or Assam Himalaya. 2. The Central or Nepal Himalaya. 3. The Kumaon or Western Himalaya. 4. The North-west or Panjab Himalaya. The Western Himalaya extends for about 320 kms between the river Kali and the great basin of Satluj. All the eight districts of Kumaon and Garhwal lie within this region. As such this Himalayan part can also be termed as Uttarakhand Himalayas and Garhwal Himalaya is a part of this division, restricted to five districts of Chamoli, Dehra Dun, Pauri, Uttarkashi, and Tehri.

Up to an altitude of 1,000m the zone is characterized by the dense forests of Shorea robusta; mixed deciduous forests with Anogeissus latifolia, Dalbergia sissoo, Lagerstroemia parviflora, and Terminalia species; swamp forests with Bischofia javanica, Carallia brachiata, Salix tetraserma, and Pyrus pashia as the chief components, with occasional cane breaks. The moist slopes are occupied by species like Cedrella toona, Eugenia jambolana and Ficus glomerata.

The isolated patches of grassland vegetation chiefly consist of Eulaliopsis binata, Heteropogon contortus, Botheriochloa *sps.,* Dichanthium *sps.,* Eragrostis *sps.,* and Chrysopogon *sps.* Above 1,000-1.200 m, usually on dry locales, Pinus roxburghii, forms excellent forests reaching up to elevations of more than 2,000 m. The drier belt, however, shows, xeric elements like Acacia spp., Mallotus philippinensis and Zizyphus jujuba, with dense patches of thorny succulents such as Euphorbia royleana on highly steep slopes. Here the grass-land patches are also frequently scattered on saddles and ridges where Arthraxon prinonodes, Arundinella setosa, Erianthus, rufipilus, Eulalia quadrinervis, Imperata cylindrica, Sporobolus diander and Themeda anathera, are the predominant grasses. In between the altitude ranges of 1,200 and 1,500 m pine is allociated with Quercus incana and Rhododendron species. Above 1,800 m, the northern slopes with moist situations are covered with oak, Rhododendron forests interspersed with Myrica farquhariana and Lyonia ovalifolia.

The vegetation at the higher altitudes exhibits a dense canopy of Quercus floribunda at moist situations, which occupies an intermediate range between Q. incana and Q. semicarpifolia. These forests are also rich in epiphytes, orchids and ferns with lianaceous climbers especially Hoeboellia latifolia, Schizandra grandiflora and Vitis spp. are common. Above 2,800m, Oak-Conifer association occurs where Quercus semicarpifolia, Abies pindrow, Rhododendron arboreum, Taxus walllichiana, Pyrus and Viburnum species are the dominant forms. One can also witness here the inner drier region in the altitude range 2,000-3,00 m with forests of Cedrus deodara. The outer range, however, exhibit pure forests of Curpressus torulosa. Nbetual utilis and Abies spectabilis occur either singly or mixed in various proportions extending up to 4,000 m. In several localities Quercus semicarpifolia-Abies pindrow forests lying south of the snows as well as the Cedrus deodara forests north of them, merge into the above type of forests.

Betula is associated with Rhododendron campanulatum at higher altitudes. In subalpine forests of Betula-Rhododendron as well as the shrubby areas of Rhododendron anthopogon, R. lepidotum, species of Cotoneaster, Lonicera, and Rosa are better developed on the north exposed slopes. In moist and open localities, luxuriant meadows of herbaceous perennials are seen with the usual components represented by general such as, Anemone, Epilobium Gentiana, Geranium and Polygonum etc. The important meadow formations are Auli and Gorson near Jamnotri, Bendi, and Kush-Kalyani etc. The curious, woody species of Saussurea are widely distributed on the moranic slopes. At several localities in Garhwal, extremely dry conditions prevail and here the prevalent plants belong to species of Berberis, Caragona, Ephedra, and Juniperus.

The richest herbaceous alpine flora is exhibited by Bhyundhar Valley of North Garhwal. The details about vegetation patterns of higher peaks of Garhwal are still unknown, nevertheless records from regions of similar elevations have been described by Swan who found fungi and yeast in soil at elevations of 7,600-8,400 m. Ecologically, this zone may act as an important area since it regulates the meteorology and hydrology of all the surrounding regions. Apart form terrestrial habitat, the Garhwal Himalayas comprise a number of lakes such as, Deoria Tal, Sahastra, Bhil Sarovar, Vasuki Tal, Gandhi Sarovar, Satopanth Lake, Hemkund, Gaurikund,

Roopkund, Tammakund, etc., roughly between the elevational ranges of 1,650 to 5,000 m. The important rivers like Alaknanda, Dhauli, Mandakani, Pindar, Nandakani, Bhilangana, Bhagirathi, Eastern and Western Nayar are additional systems where vegetation needs detailed explorations. Major biomes of Garhwal Himalayas can be classified, on the basis of the kinds of habitat, plant association and environment conditions, into the following types:

Aquatic habitat

This consists of a number of lakes at different altitudes and rivers. Plant communities occurring in this habitat can be divided into two systems:

Lentic ecosystem

Includes the habitat with standing water as different kinds, tals, and sarovars in this area, and

Lotic ecosystem

Comprising running water habitats scuh as as rivers, etc.

Terrestrial habitat

This part includes the range of mountains encompassing most important botanical regions with climate and vegetation ranging form truly montane to temperate to alpine type from low the high altitudes in successive belts. The important biomes associated with this habitat are:

The high altitude biomes

Including the regions of semi-attenuated atmosphere, low atmospheric temperature, high aridity and harsh contrast between macro and micro-climatic conditions. As such, the type includes the whole of the elevated region above the timber line on Garhwal mountains. This biome consists of:

(i) Dwarf bushy ecosystem, (ii) Herbland ecosystem, and (iii) Meadowland ecosystem

The sub-high altitude biomes

These cover the regions between the timber-line and the basal zone of conifer commencement and characterized by forests, grassland patches, as well as shrub and herb communities which can be divided into following systems:

(i) The moist woodland, (ii) The dry woodland, (iii) The grassland, (iv) The shrubland, and (v) The herbland.

The low altitude biome

This covers the regions below conifer belt including foot hills, Tarai and Bhabar and includes:

(i) The moist deciduous woodland ecosystem, (ii) The swampland ecosystem, (iii) The mixed deciduous woodland ecosystem, (v) The grassland ecosystem, (c) The shrubland ecosystem, and (vi) The herbland ecosystem.

Various factors influence the organisms in a holiastic manner, i.e., these operate in conjunction and not in isolation to each other. The different altitudinal zones of Garhwal Himalayas are chiefly affected by two categories of ecological factors:

Mani reported that Himalaya profoundly influences the climate and vegetation of the surrounding lowland region. It must be simultaneously suggested that the prevailing climate and the vegetational characters of lowland more or less greatly modify the altitude-conditioned environment of the mountain system. Truly speaking, the mountain is composed or ridges and saddles, and two opposite slopes meet either with river, rivulet or lake, Each slope of a particular altitude has its specific topography, influenced by a number of factors and the physico-chemical nature of soil is largely dependent on the parent rocks composing it. The submontane, temperate and alpine zones show irregular topography with variable soils along-with microclimatic conditions which control the distribution pattern of vegetation.

In spite of the activities of different microorganisms, animals and fire, different biomes of Garhwal Himalayas are chiefly influenced by the following activities of the natives:

(i) The domesticated animals of the natives like cow, ox, sheep, buffallow, goat, etc., damage most of the vegetation which ultimately change the soil condition. (ii) The excessive felling of different valuable and ecologically important vegetation by the natives and the forest contractors causes marked changes in biomes. Irregular felling of vegetation recently caused invasion of different weeds as *Lantana camara, Oxalis* spp., and *Tagetes* spp., etc. It is well-known that invasion of conifers in Himalayas has taken place due to irregular felling of climax loak forests:

We are making the following suggestions for checking further deterioration of the habitat:

1. Detailed ecological survey of different vegetation types in relation to their co-habitat at different altitudes to establish ecological relationship and gradient.
2. Study of phenological behaviour of different ecosystems to correlate relationships between eco-factors and vegetation in relation to physiognomy.
3. Evaluation of various plant community structures at different altitudes in a comparative manner to find out the adaptability of individuals in the community and niche differentiation; establishment of basic

parameters of different ecosystems; their structure and function.

4. Survey of scavengers and secondary producers and their role in different ecosystems.
5. Production behaviour of different communities at different altitudinal zones along with structural modifications in ecological amplitude and relation to microclimate by establishing peak growth biomass and net primary production, The study of biomass of the flora and Fauna in its entirety and inter-relationship.
6. Energy conserving efficiency and budget of different communities at different elevations.
7. Study of different communities influenced by the biotic disturbances and their relationship with undisturbed communities.
8. Determination of the ratio of each sere or segment and terminal standing crop to annual net production.
9. Experimental studies of succession by encompassing analysis of (i) litter (ii) shade (iii) shaded litter, and (iv) reduction in root competition on a regional basis in space and time. Simultaneously Continuum Index be studied for succession on the basis of relative frequency, density and abundance.
10. Observation of competition behaviour of different weeds.
11. Autecological studies of economically useful species in various directions to enable an understanding of the environmental and nutritional requirements of these plants which would help in propagating important food and fodder plants growing in this region.
12. Knowledge of steps of lttler formation and

decomposition in different biomes and its importance.

13. Study of physico-chemical nature of different soil types at different elevations and analyses for different elements in soil and plants to study the mineral cycling.
14. Effective pollution measures in different ecosystems.
15. Study of alpine ecosystems in relation to climate, flora and fauna along with its influence to lower altitudes.
16. Study of ecological approaches made by ants and spiders.
17. After evaluating and collecting the results from different biomes improvements measures can be undertaken, considering the difficulties of the natives.

The overall energy budget for Yusmarg meadow-lands indicated that 38 and 42% of total net production disappeared, respectively from fenced and grazed sites, and rest of the energy remained locked in the vegetational compartments.

The highest energy conserving efficiency values were obtained in summer in fenced and 8.23% in grazed meadowland. On the while, the energy conserving efficiency value in fenced region was about six times greater than that in grazed meadowland. Within an ecosystem the first step in the process of energy flow is the capture of solar energy by the green plants.

The productivity potential of different ecosystems, depends much on the efficiency with which the vegetation accumulates this energy in the net primary production which forms the ultimate source of the caloric needs of all the animals and men. The energy conserving efficiency of primary producers is the ration of output to input in an unit area over a definite period. The studies presently made are based on fenced and grazed meadowlands, both situated at Yusmarg.

The energy conserving efficiency has been calculated by the following formula:

Energy conserving efficiency (%) = Energy captured/ m^2/t/1/2 Solar radiation/m^2/t x 100 t = period of solar radiation and energy captured.

The net dry matter production as obtained in different months were converted into energy values by multiplying them with the caloric values of the plant material. In this investigation only half of the solar radiation was used for calculating efficiency because 50% of the total radiation is not usable by plants in photosynthesis; Terrien *et al.*, 1957; Bray, 1961. Hence efficiency of energy captured is defined here as the per cent of half the total solar radiation reflected in the total net production.

The green shoot captures the energy but the material so synthesized is distributed to other parts as well. Therefore, in order to get the energy conserving efficiency of dominant, other species and entire community, for fenced and grazed sites, the total plant biomass was taken into consideration on unit area and time basis.

The data reveal that the average caloric values were 3251.13 to 3876.69 cal g^{-1} dry weight in the fenced and 3155 to 3717.74 cal g^{-1} dry weight in herbage of the grazed meadow. Ovington *et al.* have reported the caloric values for maize field, prairie, savannah, oakwood ecosystems in Minnesota as 4525, 4127, 4817, and 4865 cal g-1 dry weight, respectively. Thus, the energy content of the present herbage seems to be lower. It is observed that the intensity of grazing influences the pattern of energy flow through the vegetation components of fenced and grazed sites. Out of the total energy fixed 42.91 Kcal m^{-2} day^{-1} is transferred to underground system in the fenced; while in grazed, 5.11 Kcal m^{-2} day^{-1} is accumulated in the aboveground parts and 5.67 Kcal m^{-2} day^{-1} in the fenced, and 0.66 and 3.88 Kcal m^{-2} day^{-1} in the grazed meadow. Singh *et al.* have summarized data from a number of grazing lands to elucidate the energy flow pattern in the vegetation compartments. Accordingly a tropical canopical grazing land ecosystem captures energy at a mean rate of 12.38 ±6.93 Kcal m^{-2} day^{-1}. Minimum energy fixation occurs in strongly water limited grazing lands of semiarid zone. During the rainy season, the energy capture efficiency is usually several times higher as compared to the post-monsoon season. Billore has noted that the values are affected with the intensity of grazing.

Humans are capable of doing much more than what they have done so far. There are many examples. Through education, training and participation in development programmes. The initial willingness and capacity of people to develop may be limited by several factors—

They make lack time or energy for cooperative or voluntary efforts. Activities required to meet immediate needs may prevent them from dealing with problems that seen less pressing. They may not expect to gain from cooperative ventures, distrust the motives of those proposing such ventures, or doubt and feasibility of usefulness of the proposed activities. They may lack the skills to correctly perform unfamiliar tasks.

To overcome these limitations—

Schedule work around people's calendar, do not expect them to work when they need to tend their fields or are busy with other tasks. Hold meetings in the evenings or when they are free even if these times are not convenient to you. Break each project into small sub-projects. The short-subprojects can be completed during the period when farmers have small amount of free time when it is easy to motivate them to participate and continue the work when they see evidence of progress. Show respect for people by arranging to have inputs needed technical assistance. By punctual and arrive at the meeting before the people are assembled.

We have proposed and promoted a seven-pronged action for people's participation in Himalayan eco-system development, these are—

Introduce re-generation and planting of fuel/fruit/multipurpose plants and trees. Revive bio-fencing, social and other fencing and promote self-generation of vegetation. Promote tree planting, develop low-cost technologies for direct seeding, production etc. Encourage local support and participation in tree-planting by increasing fast-growing multipurpose tree species; promote village Van Panchayats; increase local involvement in protecting and managing forests and woodlands; and strengthen forestry extension parallel to agricultural and rural extension services.

Harness alternative energy/mini hydel/smokeless chulhas/solar/bio-gas/wind energy, etc. Recommended actions for this programme could be recovering of logging wastes, introducing rural and urban electrification; subsidise locally available commercial fuels such as LPS, and encourage use of pressure cookers for efficient use of fuel, develop cheaper and durable substitute for fruit-packing, and provide demonstration and training facilities.

Change land-use practices to establish and rehabilitate degraded upland watersheds. Technically proven and economically sound methods are available for different social ecological conditions. Proper use can sustain and improve the productivity of both forests and agricultural lands. This programme focuses on three main activities.

1. Establishing tree and grass cover to stabilise upland areas and to ensure adequate supplies of fuelwood, fodder and building poles.
2. Controlling livestock grazing to the extent of total stall feeling.
3. Development and promoting sustainable and integrated farming systems. Areas affected by land-slides need to be urgently stabilized.

Harvest water in small-scale water dams and roof-water harvesting in ferro-cement and other tanks. Also, promote mini/macro hydral plants; and regeneration of springs through proper planting in sub-catchment.

Integrated land and water use: This will include various objects of crops, livestock, horticulture, forest, fisheries, etc. well-knitted into a complimentality that will ultimately benefit the people of the area, while preserving the eco-system. Improved land and water use will be according to the capability of the area. The objectives of the programme are to get the optimum use of each square metre of land and each cubic metre of water.

A call to the people: This is a call to the hill people who are fast shifting from their traditional hard life to one of slumber, lethargy, lack of self-reliance and indifference. The proposal includes population control, people's action groups for eco-development, environment conservation, planting, etc. The self-reliant programmes will encompass income generation programmes such as Sericulture; Apiculture; Mushroom culture; Poultry; Rabbitry; Wool and Khadi weaving, and nut culture, non-perishable fruits, etc.

In practical terms, inspiration is the ability of the project staff to give people the visions they have—what they can do to improve their situation. In this context the concept of Social Energy is important. It is known tat when people are determined and put their energy to work, insoluble problems can be solved. Social energy depends on: Idea, Ideals and Friendship. Principles and practice of people's participation have been enunciated above. There may be experiences all around but we would like to specify two such experiences for stimulating the Himalayan communities to improve their environment and also raise their potentiality.

Case studies of the Central Himalayan Rural Action Group and the KVK at the Hill Campus of G.B. Pant Agriculture University in Ranichauri have shown the results of People' participation in eco-system development. The Himalaya Surakshan Ashram with its headquarters in Ranichauri is the new addition to the band of volunteer agencies trying to help the local people towards self-reliance and economic development. Over a period of 25 years, research in the Central Himalayas of Uttar Pradesh has made me actually aware of the increasing hardship their residents have born, and the cumulative resentment they have barbored, as a result of the rapid destruction of their environment resulting from the manner of its governance and resource exploitation.

During 1981-82, a bitterly contested and heavily publicized parliamentary election took place in Garhwal, the heartland of the region, bringing to the fore conflicting expressions of regional identity and interests. At the same time there was increased regional participation in a ten year old indigenous movement to save the Himalayan environment through saving the forests upon which the indigenous people depend for their livelihoods as subsistence farmers. Militant, non-violent action to save the forests was mobilized under the rallying cry the movement has the trees" movement. These two responses in the region to its situation might also be described, respectively, as "elite politics", i.e. the politics of power, and as "grass-roots politics". i.e., the politics of survival. Although the election was confined to the tiny Garhwal Lok Sabha Constituency, it became an issue of overriding importance throughout 'Uttarakhand' and beyond. Similarly, although the activities of the *Chipko* movement took place primarily in two subregions, its goals, methods and accomplishments became watch-words throughout 'Uttarakhand' and beyond.

The Uttarakhand comprises a rectangle of about 300 by 150 kilometers, encompassing all of the Himalayas between the state of Himachal Pradesh on the west, Nepal on the east, Tibet on the north, and the plains districts of Uttar Pradesh on the south. It is made up of the eight Himalayan districts of Uttar Pradesh; "Dehar Dun, Tehri Garhwal, Uttarkashi, Garhwal, Chamoli, Naini Tal, Almora, and Othoragarh. They constitute a lightly populated, topographically rugged, relatively inaccessible mountainous appendage to the state of Uttar Pradesh. Its capital, Lucknow, lies about 500 kilometers away in the midst of the immense, flat, Gangetic plain, the western half of which makes up the bulk of the state. The people of the mountain region are predominantly Hindus, and along the northern, high-altitude fringe, some Bhotiyas.

The Hindu population includes two closely allied, land-owing high castes, the Brahmin farmers and priests, and the Rajput farmers, comprising about 8 per cent and 68 percent of the population comprising about 8 per cent and 68 per cent of the population respectively. Various artisan and service castes make up the remaining 24 per cent of the population. Before India's independence they did not own land, they were collectively known as the

Dom in most areas, and were regarded as "untouchable". The entire population except the Bhotiyas speaks what Grierson in his classic Linguistic Survey of India called he "Central Pahari" branch of the Pahari language group which is of Ido-Aryan stock and is spoken throughout the lower Himalayas from Western Kashmir across to eastern Nepal. Those who speak these languages are known conventionally and collectively as the Pahari people, and generally so designate and identify themselves, for they share many features which identify them to themselves and others as distinct and related-in short, as an ethnic group. Their languages, while mutually intelligible, are not easily understood by people of the plains.

Similarly, their marriage networks overlap throughout the hills, but overlap scarcely at all with those of plainsmen. Their culture is thus bounded and readily distinguishable form that of the plains, but is not entirely unique or divorced from it or from that of the rest of North India from whence some, and perhaps ultimately all, of their ancestors came. They share common origins. Regardless of stereotypes, however, it must be understood that the Paharis are not tribal people in any conventional or anthropological sense of the term. Rather, they are hill-dwelling.

Indo-Aryan-speaking, caste-organized, agricultural Hindus society and are well within the range of variation found in village culture and society elsewhere in North India. The degree of cultural sharing or commonality is greater within the Pahari area than between i and the plains. Theirs is not, however, a homogenous culture throughout the region. Geographic, environmental, social and political barriers of long standing have limited interaction between people, and this is most conspicuous at the sub-regional and local levels where differences occur in precisely the features shared at the broader levels : speech, ceremonial forms, deities worshipped architecture, dress, ornament, folklore, technology, organization of Himalayas, and within each of their major regions, and again within their constituent sub-regions, down to particular water-sheds, valleys, hillsides and localities, are social, cultural and linguistic territories of varying degrees of difference and inclusiveness which are recognized and acted upon, especially to the interaction and cultural commonality within these area translate under circumstances of confrontation with people of contrasting characteristics and interests.

There is, in effect, a system of regional or segmental ethnic identification analogous to the system of segmental lineage organization found within each caste of the region and of India, and well-known elsewhere through the literature of social anthropology. Just as the shared and distinctive characteristics of Pahari culture may obscure is subregional and local variety to the outsider, so its subregional cultural and social differences may obscure the common culture to the people themselves. This is no doubt a feature of all ethnic identity. A look at any contemporary political map of South Asia will reveal an important feature of the political context for regionalism in the Central Himalayas.

Most sectors of the Himalayas have attained a degree of self-governance, a reflection of the topographic and ecological distinctiveness of the environment, the socio-cultural distinctiveness of the population, and the assertiveness of the mountain peoples. The socio-cultural distinctiveness has been the basis peoples, while the distinctive environment has provided the material conditions upon which the common interest of the mountain populations are based, recognized and asserted. Together, these facts have focused the attentions and energies of the Himalayan peoples towards improving their situation through self-governance.

The politics of the Uttarakhand reflect the minority status of its people, comprising as they do but 4 per cent of a population of 110,000,000 in Uttar Pradesh, a large number of population not much aware of the culture and environment of the mountain region. The politics of the Uttarakhand also reflect the fact that it constitutes of precipitous, fragile mountain land, rich in forest, mineral ad hydro-electric resources in a state of 300000 sq.km. of alluvial land. The desire in the Himalayas for regional self-determination and the desire in the plains for the Himalayan resources, with the added fear of separatism and fission, create a mutually reinforcing system of contradiction, friction and provocation.

When an extensive network of roads was built throughout 'Uttarakhand' following the Indian-Chinese border conflict of 1962, the motive was clearly strategic, but a major consequence was the opening of this formerly largely impenetrable area, rich in resources, to traffic of aii kinds. Timber and other forest products ranging from resin to medicinal herbs, and minerals ranging from limestone to potassium and rare metals, suddenly became accessible to contractors and other entrepreneurs. The blasting of laborers used in road-building, were soon replaced by those used in extracting resources.

The overall environmental degradation wrought by road construction-the massive erosion and landslides caused by the road cuts and blast-shocks, the resulting loss of soil, forests and water sources, and the decimation of firewood and other forest products by labour crews and others together with the social and economic dislocation those brought to the local populace-was quickly awarded by the same phenomena on a magnified scale as a result of the extractive industries' activities. Hydro-electric sites along the Ganga and the Yamuna rivers and their tributaries were promptly exploited with similar consequences. Now, for example, the Tehri Dam, Asia's largest is being built on the Bhagarathi where it will submerge historic Tehri City and many villages. In addition, when the roads came, land previously cultivated by the local farmers for their subsistence crops or devoted to pasturage, suddenly also became of interest to entrepreneurs and apricots, and other commercially valuable crops suited to the environment such as potatoes, ginger and opium.

Industrial giants vied with lesser capitalists in buying up and exploiting the new found agricultural bonanza at the expense of local people. The Himalayan shrines of Badrinath, Kedarnath, Gangotri, Yamnotri and other places sacred to Hindus became accessible to pilgrims and tourists by bus and taxi in a day or two of travel rather than the weeks or months of trekking formerly required. This new form of mass tourism taxed the capacity of the Himalayan environment heavily as it populated the pilgrimage routes with hotels, restaurants, shops and other impediments of the industry.

At the same time, and on many of the same routes secular tourism flourished, attracting plains-men and foreigners alike to the vacation spots featuring snow-views and wildflowers, touring, trekking, hunting fishing and mountain climbing. Tens of thousands annually take advantage of these attractions with little or no attention paid to their impact on the environment of the mountain people. The major beneficiaries of these kinds of development are outside entrepreneurs, their customers and their patrons. Most of them are absentee owners from the plains, a few are expatriate mountain people, and a very few indeed are elite, educated, wealthy, plains-oriented residents of the mountains.

On the whole, local people are not properly employed in the enterprises which the opening and "development" of the mountain regions has brought. The usual impact of some of these development activities has partially been to deplete the forests, erode the soil, dry up the water sources, pre-empt the firewood, fodder and building materials, co-opt or destroy some of the viable agricultural land and pasturage, leaving many of the mountain people unable to make an adequate living.

As a result of this and other reasons, in the past 20 years, there has been accelerated emigration of men to the plains in seek of work, most of whom become part of the oversupply of urban unskilled, underemployed. This has resulted in Uttarakhand villages which have been described by some observers as "maleless", where, as in one described by Rawat, cited by Rao: most of the adult and able males have emigrated leaving behind their wives, children and old parents". These people are left trying to run farms with broken families, partially depleted soil, reduced forests, receding sources of firewood and fodder such that required daily supplies are up to half a day's walk from the villages, and diminishing streams and springs such that water must be carried for up to an hour several times a day. The effect on Uttarakhand is not lost on observers of the Himalayan scene. Dhasmana, in a paper on "Transformation an Development in Uttarakhand," observes that.

> "Buy the grace of holy Ganges the hill districts of Uttarakhand are tied with the plains of Uttar Pradesh. Part of an over populated state with more than fifty districts and a state capital situated more than five hundred kilometers, the hill region view of its histo-geographic past requires special attention".

One of the most noted foreign experts on the Himalayas, German economist Hans Rieger, says detailed the decimation of the Himalayas with these remarks:

> If this pace of progress goes unchecked and

> unplanned is every danger that the very existence of life dependent on them, directly or indirectly, may be in jeopardy. It is now high time that the Himalayan states may form a consortium for the protection, preservation and conservation of he natural resources of the Himalaya".

Bharati has observed that "Himalaya is a state of mind," that they "tend to be ascriptive rather than actual mountains." He goes on to quote T.V.R. Murti, noted Hindu philosopher: "Existentialism says that poentiality is more important then actuality. This is true: the Himalayas of the rishis and the yogis and the miserable huts of the people there". The relevance of those statements in this context is that the Himalayas, and particularly that portion called Uttarakhand constitute the Holy Land of Hinduism-in a sense, heaven itself.

Like all holy lands and heavens, they are idealized. They are the abode of the gods, the place of celestial beauty and perfection. Fonia says, in his book, *Uttarakhand: Garhwal Himalayas,* Perfection is the other name for this holiest of the Holly Land." The Hindu scriptures repeatedly praise the glory of the Himalayas: "In a hundred ages of the Gods I could not tell thee of the glories of Himachal. As the dew is dried up by the morning sun, so are the sins of mankind by the sight of Himachal". But unlike many holy lands and most heavens, the Himalayas have a solid tangible reality and flesh and blood people make their living there through the dint of very real sweat, employing easily observable customs, traditions and practices, according to ideologies which they can readily articulate, and under eminently tangible material, political and historical circumstances.

When outsiders and idealists are confronted with the disparities between the ideal and the actual Himalayas, and especially between ideal and actual Himalayan peoples and cultures, the result is likely to be traumatic. For the people turn out to be hard-working farmers, herdsmen and artisans, rather than saints, rishis or yogis, much less gods. Their culture turns out to be pragmatic and mundane rather than other-worldly and idyllic. Their beliefs and customs prove to represent simply another in the wide range of sincere but variable interpretations of the traditions that comprise the whole of Hinduism, rather than meeting the strictures of the most demanding Shanskritic texts and the most pious of Brahmanic advocates.

> "There is the pervasive feeling among the learned in the Hindu tradition, that ideologically important concepts cannot be expressed except in elegant or at least standard grammatical Sanskrit. Meat eating, irregular bathing, unpredictable sex habits, and bad or no Sanskrit in religious matters form a quartet in the Brahmin's mind...a `noxious fusion'.."

Yet mountain people exhibit these very characteristics, unholy in the eyes of some plainsmen, while they occupy the holiest of environments. If outside advocates of orthodox Brahmanism are alarmed at these revelations, a section of the educated elites of the mountains themselves, and especially those who are expatriates, are embarrassed as well as alarmed, for they are anxious to the identified with the ideal rather than the actual Himalayan culture. They are likely to the people who have denied, repressed, forgotten, or concealed the actual in favour of the ideal, in confidence that the later-and the cultural mobility it represents is the avenue to the social, economic, and political mobility elsewhere described this as "plainsward mobility," or "plains emulation".

Another quotation from Dhasmana will help to define the situation. He identified what he calls the "migrant syndrome" times". He continues, describing it in these words:

> Whatever may be the reasons of origin of this

> characteristic pattern and without challenging its authenticity or questioning the loyalty of its people for their land, it was to be immigrants...This syndrome as a status symbol was in vogue for over four hundred years. During this period leading class of Brahmans and Rajput castes could list their claims to have migrated for different parts of the country and established a superior hierarchy over the remaining people..The migrant claim was a race for Himalayan culture. In the process the people living in their own area developed contempt for their culture and ethnic origin. It divided the population into various hierarchies, stratified social mobility..and hindered socio-economic progress".

He goes on to report that under British rule, form the mid-nineteenth century, "The people of Uttarakhand started trickling to plains for education and in search of livelihood," returning seasonally, or at retirement, or not at all, depending upon their experiences outside.

> "But the Garhwalis who attained some success or economic status took their families with them and started setting down in urban cities where they worked. Very few of them revealed their origin. A great number of such settlers dropped their clan titles and adopted the common prevalent caste titles in plains".

But in the past 30 years, another change has occurred, he notes. Many emigres have fared well in the cities of the plains.

> "They are not shy of their origin and are proud to call themselves Garhwalis. People who had dropped their clan titles have reverted to their usage. But this attitude has not been of any help to the region of their origin. After serving in plains the majority of the hill people are not returning to their place of origin. A new concept in the migrant syndrome has become of the status symbol today. This is the concept of being a Garhwali but settled outside

> Garhwal. Ninety-nine per cent of the Garhwalis who attain success or economic status settle down in urban towns. consequently the high-land social system is crumbling. Its culture is disintegrating. The new generation of Garhwali settlers are devoid of their language, rituals and social customs"

The actual Himalayan culture becomes an embarrassment to those who aspire to the definitions of respectability demanded by the Brahmanic plains culture, for they have adopted plains standards as their own, in a typical instance of adoption of an alien reference group.

Newcomb, alien reference group. Newcomb, 1950: Plainsmen for them have become that type of reference group that has termed an "identification group", "the source of the individual's major perspectives and values," whether or not one is recognized by others, or even y oneself, as being a member of the group. The culture of one's membership group may be more than an embarrassment when it deviates from that of the identification group, for it becomes an obstacle to mobility and the advantages that promises. The response by those who take offence at the actualities of mountain society and culture is varied, but all to one end to disassociate themselves from these attributes and from any hint of legitimizing them. To do, so they must also disassociate their compatriots, their caste and their religion from the compromising attributes, their caste and their religion from the compromising attributes. 'Therefore, they simply deny that mountain people of their own ethnicity and region are less than paragons of Shanskritic, Brahmanic plains-style orthodoxy in culture, social organization and even language.

The vehemence of the disassociation from actual Himalayan culture, and the cagerness to identify with an idealized one, is indicative of the emotional charge associated with the issue. Certain tactics are recurrently employed by those faced with reports of customs, beliefs, practices, norms and social organizational features contrary to those envisioned as embodying the ideal Himalayan culture. Among them are the following:

(1) Reports contrary to the ideal are sometimes branded as false, and often as defamatory as well. They are attributed to gullibility, misinformation, misinterpretation or deliberate misrepresentation.

(2) If this is not plausible or palatable, then it is likely to be asserted that the reports were in fact derived from regions, peoples, or ethnic groups other than the one reported or other than the one with which the respondent identifiers. That is, it is charged that the characteristics reported were infact observed not among Hindus, but among "tribals", or not among "caste" Hindus but among untouchables, or not among people of Uttarakhand but among people of neighbouring and suspect regions, or not among genuine Garhwalis, but instead they must have been made among doubtful, peripheral Garhwalis. In short, the reported observations, if true, were made not among "us," but among "them".

(3) Resort to the categorical local designation "tribal" is a particularly widely and effectively employed defence in these circumstances, useful because it has little explicit, substantive meaning. It is simply a category used to distance, exempt or exclude any group at all which manifests "actual" Himalayan culture from those claiming to manifest the "ideal" version, or to exclude any socio-cultural feature regarded as backward, exotic, alien or simply different. It is condescending but it can be used to imply sympathetic concern for the well-being of the group in question by suggesting that its improvement by undertaken while at the same time some distance is being placed between that group and ones own.

The "tribal" designation thus separates and exempts exactly where the term "Paharj" unifies and includes. There is no limit on the application of "tribal" to exclude those who do not make one's own claims, or who do not believe or behave as one thinks they should.

Complementarily, there is no escape from applicability of "Pahari" among those who live or hail from the mountains. Doubtless this explains why some mountain elites and expartriates resent and deny the designation "Pahari". It prevents them from disassociating themselves from the actual world of their country, cousins, neighbours. This is the source of the vehemence of their objection to the term: as a collective description of all Himalayan Hindus, it ignores the very distinctions upon which their claims to ethnic superiority are based-distinctions of sub-region, sub-culture and caste.

The technique of exemption, exclusion and self-elevation described above, are limited almost entirely to some elites in defence of their claims to superior socio-cultural identity. Rural village folk continue o value their "actual" culture and social organization as they pursue their livelihoods according to the tradition and distinctions to which they are heir, as those traditions have been handed down and modified over the generations, serving them well in their unique and challenging environment. They honour their traditions and practise them with pride. They do so, that is, unless and until they are resocialized by outside residence and schooling, or they migrate leaving their land and livelihood so that they are unable to continue to survive according to the traditional ways.

Then do they accept to the definition of progress implemented by plainsmen and elites. Those who are forced or enticed into changing-into giving up their traditional ways to become part of "modern" plains-oriented society and of the dominant political and economic systems of the state and nation-often in that process give up not only their traditional ways of life and their means of livelihood, but sometimes their homeland too. Thus there has merged a split in the identity of the people of the Uttarakhand, corresponding to ideal and actual definitions of their socio-cultural situation.

The ideal identity is that claimed and imposed by some alienated, urban and expatriate identity of the mass of indigenous, rural people reflects their experience and interests as traditional mountain farmers, herds-men and artisans. This split is manifest in the contrasting interests those exemplifying these disparate identities pursue. It is expressed vividly in contemporary political and socio-economic movements and actions in the region, which we think encapsulate and highlight regional socio-cultural dynamics. The remainder of this paper deals directly with these. Electoral party politics are basically the politics of ethnic regionalism. They key to understanding them in relationship to the people of the region is the fact that are elite politics-the Grass-roots involvement is minimal. Yet the masses are inevitably involved, for they are the voters.

The parliamentary by-election in the Garhwal Lok Sabha constituency in 1981-82 pitted the dominant, national party, the Congress(I), and its candidates. C.M. Negi, against the challenge of the newly formed Democratic socialist Party under the leadership of its founder and candidate, H.N. Bahuguna, an apostate Congress(I) notable. With both candidates being native sons of the region, both promising regional benefits and development, the election focused nationwide attention in an unprecedented way on the region, its people and their interests.

Bahuguna brought about the Garhwal by-election by resigning from the congress(I) and from his seat in parliament, after a distinguished, varied and stormy career as a party stalwart for over 40 years including the Chief Ministership of Uttar Pradesh and successive cabinet Ministership under Mrs. Gandhi and under two opposition Prime Ministers, after which he had rejoined the Congress(I) as the party's Secretary-General. His resignation, on May 19, 1980 coincided with his founding of the opposition Democratic Socialist Party. The surprising feature of his resignation was not that he changed his party affiliation, but that he resigned his seat in parliament in order to give voters a chance to vote him up or down under his new banner.

The election for the vacant seat took place after more than a year, and after a vehement, controversy-ridden campaign involving personal electioneering by the Prime Minister Mrs. Gandhi, unprecedented in the remote mountain province where the Garhwal constituency is located. The campaign was accompanied by some of the most flamboyant charges, counter-charges and propaganda on record. But before the ballots could be counted, the election was challenged by Bahuguna on a variety of charges.

The election Commissioner of India declared the election void. The ballots were destroyed without being counted, and a repoll was ordered. These events, and many of those of the next few months, have been carefully researched and reported by the fact-finding committee of Citizens for Democracy. Thereafter the repoll was delayed and rescheduled for various reasons, until finally, it was scheduled for May 19, 1982-two years to the day after Bahuguna's resignation. It was scheduled so as to occur simultaneously with major state-wide Assembly elections in four other states.

The campaign was fought on several issues and at several levels. First, of course, it was fought on the issue of party loyalty and regional and national patriotism; Congress(I) announced that a vote for Congress(I) was a vote for the nation and therefore for the region, and it opposed statehood for the Khand, did not oppose it, and he did take a mildly stated position advocating the interests of his constituency, implying that what is good for the region is therefore good for the nation. Bahuguna, son of Garhwal versus C.M. Negi, also Garhwali. Third, the election was fought on the basis of promises and the region, Negi had been Minister for Hill Development in the state, but Bahuguna had founded that Ministry and created its unique budget, separate from that of the state at large, while congress(1) was industrialization of the region. So far as accomplishments were concerned, however, Bahuguna outshone his opposition with his record of having founded not only the Hill Development Ministry, but the village water scheme, the Garhwal University, and other regional projects, as well as putting his constituency "on the map" politically. It would seem that Bahuguna won primarily on a combination of personal charisma, long-standing identification with Garhwal, a record of attention to issues of concern in the region and a record of political power.

Although Bahuguna was elected, and his candidacy and its Congress(I) opposition dominated the media, it must not be overlooked that he and Negi were not alone in the election-campaign almost, but not quite alone. The Uttarakhand Revolutionary Party, was a presence in the Garhwal by-election although it did not run a candidate. Its platform was much milder than the party's name might suggest, but it was of special interest because it advocated that Uttarakhand be constituted as a state within the Indian republic. Aside from that, it advocated that in this way the deficiencies, problems and evils of the region and its culture in the relation to neighbouring regions could be corrected, the outflow of its resources stopped, and the region developed quickly to bring it to a par with other regions. The party has little popular support beyond the small group of intellectual and professionals who are is spokesmen and leaders, and it receives little public attention. One Dependent candidate did enter the fray, whose platform was nearly identical to that of the UKD, but he received only three tenths of one per cent of the vote.

There other Independents received comparable votes. In general, neither the campaign statements nor the votes received by Independent candidates were reported in the press. It was therefore virtually a two candidate election. Finally, there was also a Community Party presence reported as well-more prominent in the news than the UKD or the Independents, but with no candidates espousing is position. It principle manifestation was in the form of graffiti attributed to it, and the reported existence of some students, organized labour and labour organizers advocating is programms, especially around major construction sites in the region such as thaw of the Tehri Dam.

The appeal of the Communist Party in Garhwal is reported to be primarily to students and labourers. On the other hand, people associated with it have been represented activity among such rural, grass-roots activities as social work, organization of labour and co-operative, and environmental preservation efforts. This is the only political party in Uttarakhand, that we know of, which explicitly participates in such activities.

Party politics as exemplified in this election constitute an overwhelmingly elite process. Some of is candidates, proponents and active participants are elites. Some of its methods are those which reach elites. Some of its programs are those which appeal to elites. Some of its claimed achievements are those which gratify and benefit elites. All of these benefit massively the very few people of the region already involved in the national economic and political arena in a profitable way.

Politics is by definition power brokerage, and party politics is power brokerage on a regional and national scale that goes far beyond the mountains. The socio-cultural impact is equally oriented to the benefit of elites. Party politics seeks to identify the regional culture and society according to the elite, idealized view of those who participate in the benefit from those politics. It promotes a view of the mountain people and heir culture as one of unique Hindu orthodoxy.

An assessment of geological resources of the Himalaya, is a necessary prerequisite for development of the areas. These would call for in-depth study of the soil, mineral, water and forest resources, etc. Man's latest knowledge should be pressed into such an assessment. Geological factors can introduce or remove the constraints on development and their study, with necessary inputs and appropriate government support, is therefore essential for planning, so as to improve production, without disturbing seriously the ecological system.

The crucial element in an ecosystem is to ensure against undesirable violations of picturesque mountain contours, erosion of rocks and oil and also to provide properly the means of regeneration of natural resources as well as to maintain the colour and fascination of the Himalaya which is rich in scenery, minerals and herbs etc. Depletion of the ecosystem due to ill-planned schemes can deprive mankind of most of its surveys of Himalayan resources, scientific information available to a geologist would certainly be most useful. A full-fledged

Himalayan geology cell should be located in the Planning Commission, to have development plans of the Himalaya scrutinized at greater depth, from the point of view of minimizing potential geological obstructions in the implementation of schemes. Generally the agencies concerned with the construction of roads and communications, do not sufficiently take into account the effect of their plans on mountain sites, horticulture, herbal wealth, possibilities restoration of ecological balance to the extent possible. This happens, because sufficient liaison is not ensured between the geologists and these agencies. For instance, schemes of road building, deforestation, etc. should be undertaken only along with the basic and matching measures for ecological security. This is not possible without taking the planners into confidence with the geologists.

Very often important geological and location difficulties are not adequately considered in planning development projects for the Himalayan regions. This was been particularly the case in respect of the plans for the Tehri Garhwal-Kumaon region. Following points, predominantly geological, but otherwise also obvious, should be kept in mind.

Almost all forms of the land use, including those which are least affected by erosion need to be rehabilitated. Their rehabilitation would, to some extent, check erosion and floods in the Himalayan areas. While planning or suggesting the ways and means for the land utilisation proper geological factors should, therefore, be taken into consideration at the planning stage itself. The qualify and the type of soil depends mainly on the types of rocks found around the area. Wherever sandstones and quartzites are found, the soil of the region will be sandy and not so much capable of absorbing water. However where the soil is formed due to disintegration and decay of shale, slate, phyllite etc. it is often fine-grained and can soak water to a comparatively greater extent than the sandy soil.

A soil of a limestone terrain is not fertile, due to high concentration of lime. These and other varieties of soils and rocks, more studies of geological factors are called for, but this aspect is often overlooked. If land forms of a region are carefully studied and the results placed before the planners, the latter would be able to assess fertility of the soil and potential yield of crops, type of requisite fertilizer inputs etc. more accurately. Mere use of improved variety of seeds, suitable for plants with fertilizers, will not really help to secure an increased yield of crops in the mountain region.

Climatic and ecological factors differ vastly in different parts of the Himalayan region. Proper assessment of geological aspects, would avoid wrong inputs of seeds, fertilizers, etc. and make the schemes realistic and consistent with actual needs of the region. Recently, the U.P. Government installed electrically operated pumps in the lower regions of Bhagirathi valley, at Padiyargaon, to throw up the water, for inundating the higher levels, in that limited area, to make the soil more productive. If such schemes are really commercially viable, they should be extended, wherever feasible. It is not known whether adequate consultation with geologists was made to determine the location and coverage of such schemes in order to benefit maximum areas in the Himalaya.

Drinking water is the basis of life and the related schemes have to be planned, keeping in view, optimum benefit to the affected populace. Garhwal and Kumaun regions have been particularly unfortunate as amenities of potale drinking water are very meagre in many areas, though the rivers and streams flowing from Garhwal and Kumaun provide an extensive source to the people in plains for making potable drinking water from them through various development schemes.

Many areas in the Garhwal and the Kumaun region are endowed with natural resources but cannot be developed because of lack of drinking water facilities. In the circumstances, all schemes of providing potable water in this region would full a pressing need. Greater association of geologists in the feasibility studies, formulation and implementation of economic projects is desirable. It is also seen, that assistance by Block Department Officers is given to farmers on an ad hoc basis, without considering in terrain of fields, or whether geological factors would allow such assistance to be of any use. The assistance is frittered away in unproductive ventures. Without due consideration of geological limitation, money would continue to be wasted. Greater association of geologists and other administrative agencies can avoid many such pitfalls.

Improper utilization of water resources may adversely affect the ecosystems of a region and may do untold harm to the nation. There is undue concentration of dams construction already constructed or proposed in the Garhwal Himalayan region. These refer to 23 schemes—1. Ramganga Dam, 2. Ichari Dam, 3. Kishau Dam, 4. Chibbro Power House, 5. Lakhwar Dam, 6. Biyasi Dam, 7. Dak Pathar Barrage, 8. Khodari Power House, 9. Manseri Dam, 10. Harsil Dam, 11. Loharingag Barrage, 12. Pala Dam, 13. Tehri Dam, 14. Koteshwar Dam, 15. Kotli Bhel Dam, 16. Rishikesh-Hardwar Hydel Scheme, 17. Utyasu Dam, 18. Hanumanchatti Barrage, 19. Tapoban-Helang Hydel Scheke, 20. Dhauli Ganga Dam, 21. Gauri Ganga Dam, 22. Jamrani Dam, 23. Pancheshwar Dam.

The Tehri Dam will be 580 feet high rock-fill structure and will be among five highest dam of the world, designed to produce 600 megawatts of electricity and to irrigate 164,000 acres of land. Superficially, the gigantic scheme appears to be most attractive and if only national pride were to be served, any gigantic scheme would be welcome indeed. However, this particular scheme is likely to damage the ecological aspects of the region.

The scheme would, besides, involve total displacement of about 80,000 persons and loss of herbal, forest, mineral and natural wealth of untold quantum on a massive scale. This wanton destruction may be nugatory of the benefits that the scheme is likely to yield; the losses are too prohibiting to be brushed aside lightly. According to Shri S.P. Nautiyal, the rocks exposed around the Dam site are highly contorted and sheared and are not suitable for the proposed 850 feet high concrete dam. The hills above the dam site are also constituted by highly deformed and soft rocks, which may slump down in the reservoir.

The dam proposed is surrounded on three sides by major fracture zone, also known as the Dharkot thrust, and is an equivalent of tons Thrust. This folded thrust in the southern part of Tehri Dam is known as the Krol Thrust. Several Tear faults with a NWSE trend exist at Tehri Dam region and its adjoining parts and the earthquakes of great intensities are frequently generated along such faults. Similar tear faults are met with at the Marora Dam site investigated in 1945-1951 and the project had to be abandoned.

Keeping in view these technical difficulties for a concrete dam, proposal for a rock-fill dam has been suggested, but in the opinion of Nautiyal there is a great dearth of suitable rock-fill material in the vicinity and there is no outcrop that may yield the rock-fill material in large quantities required for the dam. Further, the site is very fight and narrow and poses constructional hazards for a rock-fill dam.

Comparing the rate of siltation of the Sutlej with the Bhagirathi may bring many times much greater silt than the flat gradient Sutlej. This is due to the fact that the Sutlej has not large glaciers in its course, while the Bhagirathi and Bhilangna rivers drain the southern slopes of the central snow-covered Himalaya. There had been a major blocade of the Bhagirathi in its upper reaches in recent past. This had been ascribed to some recent tectonic activity along a major fracture zone exposed at Joti and Gangnani villages in the higher parts of Uttarkashi district, as a result of which unusual devastating floods were witnessed by the river Bhagirathi/Ganga in 1978 and this posed a serious threat to Uttarkashi and Tehri towns as well. As such, the expected life of the Tehri reservoir may be only of 30-40 years instead of the assumed 100 years. Indeed, Tehri Dam is a risky venture and in the event of failure of the rockfill structure the Tehri reservoir will cause floods in the Ganga river, rising over 200 feet from the present level. Incidentally, it may be remarked that the geologists who approved the Maneri Dam of the Bhagirathi/Ganga Valley have proved to be wrong since one of the tunnel of the already constructed structure developed a major crack in 1978. It is certain, that the standfill of the proposed reservoir measuring 45 sq. km. will get polluted due to damming of the Bhagirathi at Tehri. It would adversely affect the ecosystem of the region.

Ecologists have called the Himalaya an endangered mountain. Its forests have already become a ghost of their lush past. There would be extensive felling of trees in connection with Tehri dam construction also for the settlement of the displaced people. Many pockets of forests, consisting of Tun, Rhair and Shisham occur along the Bhagirathi valley, specially in the region of the proposed reservoir site. All these commercial forests would be submerged in the waters of the reservoir and will add to the imbalance of the ecosystem of the Tehri Garhwal Himalayan region.

In extensive areas of Kashmir with natural pastures, varying from grasslands and woodlands to grass steppes and dwarf shrub alpine steppes, the main source of subsistence is domestic livestock husbandry. Although this type of husbandry occupies an important place in the economy of Kashmir, no progress has been made in formulating scientifically based management programmes. The grazing lands have been divided into alpine and subalpine, forest, and low-land pastures. These grazing lands differ in the vegetation composition as well as in available fodder. The forest grazing lands are the most important productive pastures in Kashmir. It has been estimated that in Jammu & Kashmir State the percentage of forest going cattle in relation to total bovine population is 113.5%.

The swamps, perennial depressions, and the fallow-lands are the valley grazing lands. The yield of the Himalayan pasture land is generally low because of the degradation of the pastures. Due to indiscriminate grazing and absence of grazing regulations, several pastures are in a bad condition and there is danger of extermination of a number of species of higher fodder value.

A regional project for the evaluation of grass-land resources is highly essential. The objectives of the project should be : collection of basic information on distribution and characteristics of grasslands, identification of problem in optimal use of the resources, establishment of priorities for research and rapid surveys of grazing land resources in the State. In extensive areas of Kashmir with natural grazing lands varying from grasslands and woodlands to grass steppes and dwarf shrub alpine steppes, the main source of subsistence is domestic livestock husbandry. In spite of the fact that this type of husbandry occupies an important place in the economy of Kashmir, no progress has been made in formulating scientifically based management programmes.

The management programmes which should be based on the recognition of the various ecosystems and the floristic structures in the area as the structural entities of the landscape. It has been recognized that he plants are excellent indicators of habitat factors and since vegetation forms an integral and characteristic part of an ecosystem, the area occupied by an ecosystem can be delimited by its plant community. Apart from scattered reference no work on applied phytosociology has been done in this area. In spite of the fact that there are available large tracts of grasslands in the area, the fodder resources are inadequate, because of the lack of proper management research and, probably, because of the increasing bovine population. This necessitates a properly organized research programme in pasture management, fodder cultivation, and pasture production.

The climate of the valley is sub-Mediterranean and is characterized by precipitation concentrated in the cool season with short days, where the summers are hot and dry. The total precipitation varies from 500-700 mm of which about 40% is received in the form of snow during winter. The lowest mean temperature during winter is -2°C to -8°C and the highest mean temperature of 32°C is touched in summer. The relative humidity ranges from 4090% during the year.

The climatic parameters fluctuate considerably from year to year, as well as from lower to higher elevations. The growing season in the upper reaches is up to 170 days, from May to October. The variability in precipitation affects the growth of grassland species. The clover develops only in the cool and temperate bioclimates. The 5°C is the cold limit for temperate pastures of Setaria, Paspalum, Digitaria, Pennisetum, etc.

The length of the growing is important for assessing production. The length of the productive season depends on available precipitation, temperature and the capacity of the soil to store and to release water. In general, the following conditions are recognized.

Alpine		
Cool winters	5-6 months growth	2 rest periods (Summer and Winter)
Cold winter	4-5 months growth	2 rest periods
Subalpine		
Cool winters	6-7 months growth	2 rest periods
Cold winters	5-6 months growth	2 rest periods
Temperate		
Cool winters	7-9 months growth	2 rest periods
Cold winters	6-7 months growth	2 rest periods

A classification of plant communities can be based on three criteria: physiographic, physiognomic or floristic characters; only physiognomic and floristic criteria are suitable for grazing lands. However, the physiognomic classification is often a crude approximate.

Physiognomic characters, including structural functional characters, are attributes of plants but are not definable as floristic characters. Furthermore, where fire and man play an important role in determining the structural composition of the vegetation, the physiognomic appearance at a particular site is often dependent upon the time interval since succeeding the last fire or the massive human interference. The use of floristic criteria as a basis for classification of plant communities, thus, has strong advantages as taxa are good indicators of habitat factors. In literature, the grasslands of this region have been referred to as alpine-temperate. However, the grazing lands of Kashmir are varied, numerous and little known. This diversity reflects the extreme variation of climate, altitude and terrain.

On the basis of climate the region can be divided into: alpine regions, where the climate is extremely cold during major part of the year; moist temperate region at lower reaches of the valley, between altitudes of 1500 to 3000 m; and dry temperate regions, the Kashmir valley proper and the mountains on its north-eastern side. However, for the present the grazing lands are classified on the basis of altitude, plant association, and configuration.

Generally the pastures dealt with are meadows, glades and steppes, ranging from 3000 to 5000 m in altitude. The alpine pastures are covered with snow for the most part of the year and are open to grazing from June onwards till late September. These pastures are utilizing by the truly nomadic migratory flocks and herds, which have fixed centre and follow seasonal routes to grazing areas. The valleys are usually surrounded by precipitous mountains which are covered with *Betula utilis, dwarf Salix* spp, and *Juniperus* spp.

The valleys are drained by glacial streams and support a rich herbaceous vegetation. Two grassland types can be recognized on the basis of soil moisture and depth. The eroded slopes have Puccinella spicatum, Poa oldine, P. cratensis, and Heïrochloa laxa. On well drained slopes and flats of meadows with deep soil and moisture, phleum alpinum type occurs with Phleum alpinum, Agropyron semicostatum, Poa alpina, P. annua, Calamogrostis littorea and Bothriochloa pertusa. On the basis of species content, the pastures can be divided into two main 'plant associations'. The first type can be described as phytosociological unit *Poatum-phleote* in which several species occur as constants. This type is common in the well drained slopes and flats of meadows and annual *Poa* species dominate the vegetation. The second type, which is more natural, is on the slopes, especially steppes. This type reaches equilibrium as the Puccineto-Trisetum association. This type also has several of the species as constants.

The constants of association are indicators of dry-music to mesic conditions, and several of the constants are indicators of low soil fertility. However, the constants of p-p association are mostly indicators of more mesic conditions. These can be ascribed to greater precipitation and they occur on level-ground.

Presence of constant species in the alpine-subalpine grazing lands of Kashmir

Species	*Presence %*	
	P-P	*P-T*
Poa alpina	50	5
P. Pratensis	50	2
P. Annua	40	5
Puccinellia kashmiriana	10	65
Trisetum spicatum	20	75
Heirocholoa laxa	15	60
Bothriochloa pertusa	10	40
Chrysopogon echinulatus	10	40
Cymbopogon jwarncus	8	20
Themeda anathera	5	10
Oryzops's aequiglumis	5	5
Phleum alpinum	47	-
Medicago sativa	40	2
Achillea millefolium	60	5
Carex muricata	80	-
Indicators of low fertility :		

Cerastium spp,	+	-
Koeleria cristata	-	-
Festuaca gigantea	-	+
Galium spp.	+	-
Indicators of mesic conditions:		
Artemisi moorcroftiana	-	+
Fragaria vesca	-	+
Astraglus spp.	+	+

Most of the productive pastures of the state are within the open subalpine forests. Forests provide most of the fodder requirements of the valley. These forests enclose fertile meadows, apart from the ground grasscover under the tree canopies. The land is drained by snow-fed perennial streams. There is no erosion, trappling or displacement. The litter content is moderate.

In some areas sever soil movements and erosion makes the grassland unstabilised. However, in such areas shallow gullies with stabilised bottoms are often observed. These grazing lands are visited by flocks of farmers and thus the areas have more stress and are overgrazed as compared to alpine grazing lands. It has been observed that there is always association between particular type of tree and shrub species and their ground grass-cover. However, in open meadows enclosed by forests this does not hold good. It has been found that the herbaceous species are constant in most of the forest meadows in spite of the variability in the forest type. The succession in vegetation varies in different forest type depending on the constituent taxa.

SLOPES

	Alluvial-red soils
Cutting	Pinus wallichiana
	picea smithiana
Cutting	Viburnum nervosum
	Sambucus ebulus

Grazing	Cymbopogon gryllus
	Festuca licida
Grazing	Andropogen tristis
	Dactylus glomerata
Over grazing	Stipa sibirica

The characteristic vegetation of the close and degraded forests is represented by coniferous belts. These are:

Pinus wallichiana	Juniperus squamata
Picea smithiana	J. Wallichiana
Abies pindrow	
Cedrus deodara	
Taxus baccata	

Bordering the valley at a little higher elevation than the valley proper one can see mixed forests of broad-leaved trees and conifers. The common broad-leaved tree are;

Acer caesium	Celtis caucasia
A. villosum	Alnus nitida
A. caudatum	Carylus jacquemontiana
Aesculus indica	Craetagus oxycantha
Prunus cornuta	Salix acomphylla
	S. wallichiana
Ulmus wallichiana	Populus alba
	P. ciliata

Forest degradation producers:

Viburnum foetens	Sorbus lanata
Sambucus wightiana	Cotoneaster nummularioides
Skimmia laureola	Poarrotiopsis jacquemontiana
Rhamnus prostrata	Cornus oblonga
Indigofera heterantha	Phytolacca acinosa
Astraglus grahamianus	Daphne eleoides
Desmodium tiliaefelium	Elaegnus ubmellata
Rubus pendenculosus	Berberis spp.
R. fruticosus	
Herbs	

Dactylus glomerata
Digitaria sanguinalis
Setaria italica
Danthonia jacquemontii
Koeleria crisata
Calamograstis emodensis
Festuca gigantea
F. polycolea
Bothriochloa ischaemum
OPennisetum flaccidum
Medicago sativa
M. polymorpha
Fragoria nubicola
Potentialla nepalensis
P. reptans

Swamps ad perennial depressions are frequent along the lakes in the valley and on the higher altitudes in the meadows where the lakes and mountain stream overflow. The characteristic vegetation consists of rushes:

Juncus glauca
J. maritima
J. himalensis
Carex maritima
C. curica
c. nubigene
Scirpus juncoidea
S. tripueter
Phragmites communis
Typha angustata
Sparganium erectum
Acorus calamus

Good pasture species sometimes dominate between the rushes:

Festuca polycolea
Phleum alpinum
Phalaris arundinacea
Trifolium repens
T. Pratens
Medicago sativa
Lotus corniculatus

Cropland comprise large areas of pasture lands which are usually underestimated. These are plant communities of weeds of cereal fallow as well as stubble left after harvesting. Cultivated fields are usually bad pasture lands with nearly unpalatable annual species and weeds used for 7 to 8 months, whereas stubble is used for 2 to 3 months. Throughout the valley and the slopes of the surrounding mountains there are vast paddy fields. These lands, especially, during autumn months, develop some of the best nutritious fodder plants such as:

Echinochloa crussgalli,
E. colonum
dichotma
Eleocharis palustris
E. podocarpa
E. tuberosa
Fimbristylis

The yield of the Himalayan grazing lands in this region fluctuates with the season. This is generally low because of the degradation of the grazing lands. The latter are typically underdeveloped and badly managed.

The indiscriminate grazing and absence of grazing regulations are the cause of extermination of many species of high fodder value. In early spring the pastures in good condition yield normally about 250 kg/ha. However, during summer months the yield becomes double. It has been observed that grass-legume mixtures increase the yield. For a well managed grazing land it is necessary to leave some percentage of production as unused. This is called 'carry-over'. This carry-over takes part in subsequent reproductive process and the plants remain in a healthy condition. Usually 45% of the production should be carried over and 55% should be used. In the pastures the plant communities are usually composed of some grasses, few sedges and some palatable herbs which contribute 90 to 95% of the total production. In table 2 the common species of fodder value, their production, carrying capacity and palatability rating are listed.

Several factors are responsible for deterioration of good pastures. Main factors involved in the alpine and forest pastures of this region are overgrazing and absence of grazing regulations. When there is overgrazing beyond carrying capacity, the species do not produce seeds for regeneration and sometimes the soil becomes exposed, causing erosion.

A good management involves leaving behind the recommended amount of carry-over. Therefore, the area should be stocked according to the carrying capacity. Apart from overgrazing, grazing at too early stage also results in the retardation of production. Some species such as Festuca gigantea and Bromus mollis, are prevented from reproduction. This may result in the change of vegetational composition. The requirement due to rapid expansion of cattle population can be met with by increased production of forages and hence by cultivated pasture development. The exotic species have more production and generally are palatable. The cultivated pastures should have legume-grass mixtures, which not only provide the nutritional requirements but also help in the improvement of the soil condition.

The operation directed towards the improvements of the output from native grasslands, should be considered under four production levels: (1) using the resources as they stand by adjusting stock pressures and movements, selective opening and closing of watering places; (2) Improving the ground cover resources by removing competitive trees and shrubs; (3) Introducing exotic species of grasses and legumes into the existing resources; and (4) removing the natural resources and replacing them with exotic plant communities for use as pasturage or fodder. The fourth stage should be applied in selected grazing lands. However, the first step in this direction should be testing the introduced species, and selection of strains on the basis of adaptation to climate, quality, grazing resistance and longivity.

Rain forests form an important natural resources of the Indian subcontinent. This resources is shrinking at a rapid rate. Conservations are due to massive exploitation for industrial and fire wood needs, encroachment by the landless peasants, and bush-fallow rotational agriculture (jhum). The present paper considers these conservations and looks into possible remedies. The north-eastern hill region is characterized by a wide variety of climatic conditions. The natural vegetation types range from the humid sub-tropical forests and humid montane forests to alpine meadows. Much of this region receives an annual average rainfall of 2000 mm or more.

The low elevation forests of the hills and that of the Assam Valley form an important segment of the tropical rain forests of the Indian subcontinent alongwith the Western Ghat forests of the peninsular India. The high elevation sub-tropical montane forests are also rain forest types with rich species diversity forming an unique ecosystem complex which needs to be conserved. These luxuriant forest types are often supported on highly leached nutrient-deficient and acidic, soils; these systems are, therefore, fragile. Several factors have contributed to the degradation of these ecosystems. The aim of this paper is to consider these factors and suggest possible remedial measures.

The rain forests: The prime tropical rain forests of north-eastern India are distributed in the Assam valley and long the foot hill regions. In the upper Assam and Cachar hills adjoining to it are found some of the most luxuriant Dipterocarp forests of the sub-continent". In the lower Assam valley, in the Manas region in Assam and extending up to the border of West Bengal and in the adjoining foot hills of Bhutan are luxuriant mature rain forest types of rich species diversity and with none of the important species attaining complete dominance. At higher elevations are sub-tropical montane rain forests of Quercus spp., Acer laevigatum and Rhododendrons. In Meghalaya they are restricted to sacred groves (Boojh and Ramakrishnan, 1983a). Elsewhere they occur in remote areas. While these are the important climax types, there are various stages of secondary successional forests.

Mature rain forest types in north-east India with names of dominant trees

(a)	Upper Assam	Dipterocarpus pilosus, D, macrocarpus, Artocarpus chaplasha, A. integrifolia, shorea assamica and others.
(b)	Cachar Hills	Dipterocarpus turbinatus, Arto-carpus chaplasha, Amoora sp., Dysoxylon sp. and others.
(c)	Lower Assam	Pterospermum acerifolium, Mansonia dipikae, Eugenia sp; Dysoxyeun binecteriferum. Phoebe sp; and Amoora wallichii.
(d)	Lower elevation of Meghalaya	Amoora wallichii, Artocarpus chaplasha, Castanopsis indica, and Shorea robusta.
(e)	High elevation montane forests	Quercus sp; Rhododendron arboreum, Acer laevigatum, and Castanopsis sp.

Any natural forest ecosystem is subject to a wide variety of disturbance. These disturbances may be different kinds. Some of these do not necessarily arise from human activities, e.g., death of old trees, wind blow or insect pests and diseases.

Even human activities were of very minor dimensions in the past. These minor disturbances may result in gap, building and climax phases forming a mosaic such that the climax phase of the ecosystem forms a major component. Large-scale exploitation of forests by industrial man results in drastic reduction in the extent of the climax phase or even its elimination resulting in an unstable ecosystem. *Of the reported 21,040 km² of* evergreen rain forest, 8340 km^2 of semi-evergreen rain forest and 102,000 km^2 of moist deciduous forest for India, a major proportion is confined to the north-east; a narrow belt is along the western Ghats and some in the Andaman and Nicobar islands. Large-scale timber extraction has been done from this region during last few decades. In fact, a large part of the timber and plywood requirements for the country as a whole is being met with from the humid tropical forests of the north-east. According to the 1971 census data for the north-east, the land area under forest was 48%. Allowing for further exploitation of forests in the past decade and for the fact that the forest cover computed in census reports also includes various stages of degraded communities, perhaps, it would be reasonable to assume that the actual area under good forest cover is only half or two thirds of that estimated in table.

Area under forest in the north-eastern India
(1971 Census Report)

State Area (km^2)	*Land under forest*	*(%)*
Arunachal	83,578	62
Assam	78,523	36
Manipur	22,356	68
Meghalaya	22,489	38
Mizoram	21,087	44
Nagaland	16,527	17
Tripura	10,477	57
North-Eastern State (Total)	2,55,037	48

One of the methods of timber extraction is clear-felling of trees over vast tracts of land. This has adversely affected the forest ecosystem of the mature phase.

Another common practice is selective extraction of timber of large size classes of certain species over a restricted area of a few square kms as a "canopy lifting" operation, as is done in some parts of the upper and lower Assam regions. Though the intention is only to extract larger trees, in practice it almost amounts to clear-felling of forests. The larger trees as they fall down, destroying everything else on their path create much larger gaps than intended. Further, much of this operation is done without giving any consideration for management of forests on the basis of sound scientific studies.

The possible implications of this in the regeneration of trees and the consequent stability of the ecosystem on a long-term basis and the possible impact of this operation on the ecosystem function is not known. Further, since until recently, commercial operators have been interested only in a few selected species, often less than 20 out of 400 species or more present per ha, many of the remaining trees are damaged beyond recovery. The annual available timber from two of the north-eastern states of Manipur and Tripura is shown in table. While these projections give only an indication of the magnitude of timber extraction from these states and the region as a whole, no comprehensive data are available on the magnitude of damage to natural forest ecosystem types.

Annual availability of forest products from Manipur and Tripura (Units in $m^3 \times 10^3$ bamboo in $t \times 10^5$)

Product	*Manipur*	*Tripura*
Saw timber	10.0	7.1
Plywood	110.0	19.8
Pine wood	23.9	—
Hard wood for industry	147.0	—
Pine wood for industry	9.1	—
Pulp wood	—	66.0
Poles	—	1.8
Match wood	—	1.6
Bamboo	1287	5.5

Much of the forested area in the north-east is a degraded bamboo forest of various types. This is related to the shortened jhum cycle (intervening length of the fallow phase between two cropping on the same site) as shown by our studies or even reduced to a weedy community as shown through this study and other studies. Bamboo alongwith other pulp wood trees is being exported out of the region both for industrial and non-industrial uses.

The establishment of three giant paper mills in the region would create a greater demand on the bamboo resources of the region. Unless a sound scientific management policy is evolved now, the bamboo resources would be in short supply in the near future. With the three giant paper mills coming into this region and with bamboo resources dwindling rapidly during the next few years, unless well thought out strategies for regeneration and management of pulp wood trees and bamboo species are implemented soon, pressure on hard wood species from the tropical moist forests would increase. This is particularly so when forest technology has recently developed methods for pulping tropical hard woods in a single mixture.

Much of the firewood consumed in this region, however, comes from secondary forests. Only a small part comes from primary rain forests. In India, it has been found that from the peripheral zones of the forest boundaries about 70% of the firewood requirements are met with, and beyond 10 km the use of firewood steadily decreases until at about 15 km it is almost nil.

The increased population pressure and rapid depletion of primary forest cover in the region is rapidly changing this situation. The rural tribal population of the north-east are chiefly dependent upon fuel wood for energy for cooking though other items such as rice husk may also be used. The fire wood consumption pattern in Sethliew, a small village in the Khasi Hills is given in table 4. In this village of 20 members, 3060 kg of fire wood (52455 MJ of energy) is imported each year, accounting, i.e., 5033 kg fuel wood per year. While rice husk is used as a source of energy here, there are areas where this may not be used and the villagers are exclusively dependent upon fuel wood. On the other hand, there are also areas where fossil fuel in the form of coal is used.

Annual fuel consumption for cooking in the Syth Liew Village ecosystem

Category	*Quantity used*	
Energy equivalent		
(kg)	*(MJ)*	
Fire wood	5033	86276
Rice husk	328	6032
Total	92308	(92033)
Fire wood imported	3060	52455

The dependence of the rural population on adjacent forests for fire wood is substantial. Consequently, there has been a fast depletion in the fuel wood resources of the region. Methods that increase the efficiency of utilization of fire wood should be employed, for it is low in the developing world. Here the per capita consumption of energy for cooking is considered to be between 2 to 2.5 time more than in the west.

If about 5000 kg of fuel wood were considered to be the average annual consumption by 20 individuals, according to our Sethliew village study, the annual consumption of fuel wood in the villages for a population of 19.58 million (according to 1971 census) would work out to about 489 X 10^7 kg. However, it should be realized that the population has gone up by 30% or more during the past decade and that the fuel wood resources have simultaneously gone down drastically resulting in shortages and high cost for this resources. In order to meet their fuel wood requirements, the villagers are now forced to go out of the village boundary up to a distance of about 10-15 km.

The conversion of rain forests to commercial tea plantation in Upper Assam valley and in the adjoining Cachar Hills was one of the major agricultural operations which started in 1834. In 1961, the area converted to tea was about 178×10^3 ha and by 1978 the area was about 250×10^3 ha. This accounts for about 53% of the total area under tea plantation in the country.

Forest farming variously termed as shifting agriculture, slash and burn agriculture or rotational bush fallow agriculture (locally called jhum), as practiced in the north-east India is described in detail elsewhere. This land use is a well known practice in the rain forest areas and has survived for thousands of years with little damage to the forest ecosystem. In fact, this farming practice with a long intervening fallow period between two successive cropping (henceforth referred to as jhum cycle) of 20-30 years or more has been in harmony with the environment giving reasonable economic returns of varied kinds (cereals, legumes, vegetables and even fibres) to the farmer.

Such a long cycle had maintained a balance between the gap, building and mature phases of the rain forest ecosystem. However, the long jhum cycle was possible partly because there were few farmers per square kilometer, mostly less than 5. In recent times, however, the pressure on the land has increased many-fold resulting in much shorter jhum cycles of 4-5 years. This in turn has increased the land area under the gap and building phases at the expenses of the mature phase thus, upsetting the ecosystem balance.

If one looks at the population trends in the north-astern hill states, one finds that in most of the states, with the exception of Tripura, population has slightly more than doubled in the first 50 years since the turn of the century. In Tripura during the same period it almost went up four times mainly due to influx of people from the plains. Subsequently, in most of the States, the population almost doubled in only 20 years between 1951-1971. Under these conditions, it is not difficult to visualize the impact such a large agricultural population is likely to have on the rain forests.

Area under jhum in the north-eastern India (1971 census data)

Region area jhum	Average % under area for cropping jhum/family	No. of jamilies involved		Area under cropping per year	Total under ($x10^3$)
Arunachal Pradesh	1.13	81	92	249	37
Assam	1.20	58	70	498	13
Manipur	1.20	50	60	100	60
Meghalaya	1.12	68	76	416	18
Mizoram	1.37	45	62	604	10
Nagaland	0.92	80	74	608	12
Tripura	0.51	43	22	221	10
Total	1.07	425	456	2696	17

According to Myers, there are atleast about 140 million forest farmers in the humid tropics of the world. With an average of seven members per family this would work out of 20 million families. If each family clears one additional hectare of forest each year, the total cleared area would work out to 200,000 km^2 each year. According to this estimate, perhaps half of this would be primary forest (about 1% of the total). When considered in conjunction with timber harvesting and other human activities, Myers estimates that almost 200,000 km^2 of primary humid forest is cleared every year which would work out to about 40 ha per minute. In the north-east India, jhum covered about 425 x 10^3 families according to

1971 census data. At a growth rate of 24 to 30% per decade, this now stands at about 547 x 10^3 families and is expected to go to 766 x 10^3 families by the turn of the century. If the projections given by Myers are considered for the north-eastern situation, 0.55 million families at present clear about 6000 km^3 of additional forest for jhum of which 3000 km^2 would be primary forest.

Projection of tribal families (x10^3) practicing Jhum
(After north-eastern council, 1982)

Region	1974	1984	2000
Arunachal Predesh	81	107	155
Assam	58	74	101
Manipur	50	65	92
Meghalaya	68	85	115
Mizoram	45	54	70
Nagaland	80	106	155
Tripura	43	56	78
Total	425	547	766

The exact area under jhum in the north-east is not known at present but recent efforts through satellite imagery should be helpful in this regard. However, the 1974 estimates for this region based on the total area owned by different villagers practicing jhum are given in table 6. According to this, the average jhum cycle works out to about 5 years.

Mizoram has an average of 10 yr cycle. Arunachal pradesh has been considered to have a 3 yr cycle. However, such an average is misleading as the length of the cycle is dependant upon the dispersion of the population. In the same geographical region one could find a long cycle of 30 years or a very short one of 3-5 yr too. Realizing that the data so calculated is an under-estimate of the actual area under jhum, since considerable area is desertified and is no good for cropping and this computation does not consider that rendered useless every year, the magnitude of the problem is enormous indeed.

Much can be said about the scientific basis for shifting agriculture and it is not the purpose of this essay to go into it. However, what I propose to discuss here is the feasibility of maintaining jhum in the north-eastern region under a relatively longer cycle of 10 years or more. In a study of the socio-economics of jhum at lower elevations of Meghalaya it was shown that the net gain to the farmer under a 30 yr jhum cycle is more than twice that under terrace cultivation, and that a 10 yr cycle is also economically more viable than terrace cropping under comparable situation. It was further suggested that there is much scope for more intensive valley cultivation, which is a sedentary form of agriculture with steady and sustained yield. Similar results were also obtained by Mishra and Ramakrishnan for the high elevation jhum.

The cropping pattern under jhum varies under different jhum cycles. With shortening of the jhum cycle, the emphasis by the farmer often shifts from cereals to tuber crops. Under a low fertility system as at high elevation, the shift to tuber crop, such as potato provides better returns to the farmer. This suggests that there is much scope for obtaining higher returns to the farmer by altering the cropping pattern.

The studies on energy budgets on jhum done at low elevation and at high elevation show that jhum under all cycles, both long and short, is highly energy efficient compared to terrace cultivation. Apart from the gross energy efficiency the inputs into the system also differ qualitatively. For jhum it is chiefly labour energy while terrace cultivation depends on input of fossil fuel energy in the form of inorganic fertilizers. The generally low energy efficiency of the high elevation jhum with respect to shorter jhum cycles and terrace cultivation is related to heavier input of organic manure and inorganic fertilizer into these systems which are absent at low elevation jhum systems.

Energy ratios in low elevation agricultural systems of jhum, terrace and valley cultivation (after Mishra and Ramakrishnan, 1981; Toky and Ramakrishnan, 1981)

Agro-ecosystem	Energy efficiency (output/input)
Low elevation jhum	
30 years cycle	34.1
10 year cycle	47.9
5 year cycle	46.7
High elevation jhum	
15 year cycle	25.6
10 year cycle	9.9
5 year cycle	4.6
Low elevation terrace	6.7
High elevation terrace	1.7
Low elevation valley	17.8
High elevation valley	16.2

The energy efficiency of jhum cycles cannot be considered in isolation without a discussion of the land use pattern, or else the conclusions would tend to be distorted. If land is not a limiting factor, the greater solar input into a larger area of the jhum system with a longer cycle would offset imported energy and this would ensure harmony of the long cycle with the environment, at the same time ensuring rational returns to the farmer.

With shrinkage in land area available for jhum, the longer jhum cycle can be sustained only if land available for jhum is not limiting. It should also be conceded that under different jhum cycles, the effective output per hectare (based upon the frequency of cropping after different fallow periods) would decrease due to a correction factor of 1/30, 1/15, 1/10 or 1/5 under 30, 15, 10 or 5 yr cycle. Our studies suggest that even on this basis a 10 yr jhum cycle is the most efficient in terms energy ratio and land use. Nutrient losses from the system through run-off and percolation are rather heavy during cropping; the data for N, P and K are given in table 9 for a low elevation system.

The magnitude of the losses become obvious only when the correction factor of 1,3 and 6 are applied to the three jhum cycles. Similar conclusions apply to the high elevation jhum too. Detailed studies on fertility status of the jhum plots under different jhum cycles for low elevation jhum and high elevation jhum are available. These studies suggest the generally low fertility status of jhum plots under short jhum cycles of 5 years as compared to longer cycles of 10 or more years. However, we would like to discuss at this point only the nitrogen economy under different jhum cycles at high elevation which in a large measure would also apply to other nutrients of the jhum systems.

In the present study on nitrogen budget the plots under a 5 yr cycle had the same cycle length for the past 20 years. Assuming that these plots had longer cycles prior to this 20 year period, the system has lost, in terms of initial capital, about 1.28 X 10^3 kg of nitrogen (7.68-6, 40 X 10^3kg) during the last 20 years. Thus, while 15 or 10 year jhum cycles are long enough to restore the original nitrogen status in the soil before the next cropping, it is unlikely that about 600 kg of nitrogen lost (the difference between the nitrogen capital before and after two croppings) could be restored under a 5 year cycle. One of the important disadvantages of the 5 year cycle lies in the reduced nitrogen capital with which the agroecosystem has to operate under increased frequency of fire followed by cropping with too short a jhum cycle (every 5 years under a 5 yr cycle as compared to every 10 yr under a 10 yr cycle). Our studies on nutrient cycling during fallow regrowth at low elevations and at high elevations suggest that there is a rapid transfer of nutrients from the soil pool to the vegetation compartment during the first ten years or so. It is subsequent to this that the recovery of nutrients in the soil pool takes place as a result of the transfer of nutrients through litter fall. Simultaneously substantial additions due to nitrogen fixation and mineralization also contribute to the recovery process.

The efficiency of nutrient cycling in a low elevation jhum system estimated as fractional annual turnover (the weight that left the compartment divided by the weight held in that compartment, expressed as percentage) of the different elements. While the annual percentage turnover of all elements in the soil tended to increase with fallow age, those for vegetation decreased. This is to be expected because of the marked increase in uptake by the developing vegetation and increased storage in the living plants which is higher than the rate of release through litter. The only exceptions to the pattern are for phosphrous and potassium which reached this maximum annual turnover in the soil in the 5-10 yr old fallow.

The high fractional turnover value for phosphrous in the soil in the 5 yr old fallow may be due to a higher uptake by the herbaceous vegetation that predominates there and that for potassium in the 10 yr old fallow may be related to rapid growth of bamboo (Dendrocalamus hamiltonii) at this stage, which stores much of potassium in the living biomass. A major difference between the jhum at higher elevation and that at lower elevation lies in that in the former it has to function under conditions of high soil acidity derived from the Khasi pine needles (Pinus kesiya) which also is a slow decomposer. Apart from this, the regeneration of fallow is further hampered due to lower temperature related to altitude.

The Indian woody flora is quite rich in polypetalous elements as is evident form the fact that as many as 2, 300 such species are represented in the various forest types. Of these about 160 species are the source of commercial timbers of varying importance. Anogeissus latifolia Wall., Dalberiga latifolia *Roxb., D. sissoo Roxb., Hardwickia binata Roxb., Mesua ferrea,., Shorea robusta* Gaertn.f. etc.., belonging to this group are of great importance to the Indian forestry. Since 1968, several communications have appeared on the cytology of Indian woody taxa, primarily from the Himalayan region and these have thrown considerable light on the cytological evolution in hardwoods.

A perusal of the available data reveals that many of the Himalayan woody species are still unknown chromosomally. Besides, detailed studies on the different populations have not been made in the taxa found to be variable cytologically. Further, hardly any comprehensive attempt has so far been made on the hardwoods from the Central and Southern India. Keeping these facts in view, a project was, therefore, undertaken on the cytological investigations of the Indian hardwoods. The present studies form a part of that and cover Polypetalae from the Western Himalayas. Some cultivated taxa available in the region have also been included.

Material for the present studies was collected from the wild populations growing in the Siwalik ranges and Garhwal Hills and from some cultivated individuals met with in this area. The chromosome determinations are through meiotic studies, for which appropriate-sized flower buds were fixed in Carnoy's fluid with acetic acid component saturated with ferric acetate. Squashes of pollen mother cells were made by using the standard acetocarmine technique. Desirable preparations were made permanent by processing the separated slides and cover slips through aceto-alcohol followed by alcohol grades and finally mounting in euparal. Pollen fertility, wherever referred is estimated by their capacity to stain with glycero-acetocarmine coupled with their well-filled nature. Camera lucida figures are at uniform magnification.

The cytological data on the presently investigated 61 species, falling into 21 families, in which also covers information about source of material, accession numbers and previous chromosome reports from Indian subcontinent. Only species with results of cytological interest are covered under observation.

Five shrubby species of the genus has been worked out from the subtropical and temperate forests. Meiosis is found to be normal in *Berberis asistata* DC. B. asiatica Roxb., B. chitria Lindl., B. lycium Royle and B. umbellata Wall. ex G. Don. In all these species, 14 bivalents are regularly constituted with some being of loose type. However, distribution of chromosomes during anaphase I and anaphase II, irrespective of the presence of loose bivalents, is perfectly regular. In none of the PMCs observed, any laggard is noticed either at anaphases or telophases. Even micro-sporogenesis is normal with perfectly regular tetrad formation. Whereas, in all these species some pollen malformation invariably exists, ranging from 7.4% to 32.8%. Some apparently well-filled fruits in B. asistata are found to be seedless.

It is a small shrub on rocky slopes in the Western Himalayas. Large number of PMCs from the different individuals of the species are counted to have 23 bivalents at diakinesis and metaphase I. Meiosis is regular but the pollen fertility is in the range of 78% to 84%.

Meliosma pungens walp is small to medium-sized tree in the subtropical and temperate forests of the Western Himalayas, generally in the ravines and valleys, as scattered specimens. The individuals studied from Mussoorie are found to had some B-chromosomes besides 16 bivalents at diakinesis and metaphase I . The number of B-chromosomes varies from 0 to 3 per PMC with an average frequency of 0.21. The incidence of PMCs with B-chromosomes as worked out from 300 PMCs is 18.4%. Distribution of B-chromosomes at anaphase I is quite random. Meiosis otherwise is normal and pollen fertility is almost cent per cent.

Sorbus foliolosa (Wall.) Spach. (Rosaceae) is either a large shrub or small tree, very common in the temperate forests of the Western Himalayas. The chromosome number as determined from meiotic studies on the cold temperate population from Hemkund, is 2n = 68. Meiosis is characterised by the presence of multivalents, univalents and partial pollen fertility. The average frequency pr PMC for quadrivalents, trivalents, bivalents and univalents is 7.40, 1.65, 15.70 and 2.10, respectively. In none of the PMCs. 34 bivalents could be counted. The distribution of chromosomes at anaphase I as well as anaphase II is highly irregular. Laggards which include univalents, and even bivalents, and multivalents, are encountered in 58% of the PMCs at anaphase I. Pollen fertility is low as 13.8%.

The species which is wild in some localities in the outer ranges of the Western Himalayas between 1,000-1,800 m. sometimes attains a height of a small tree. Meiotic studies in populations from the Mussoorie Hills revealed presence of B-chromosomes, besides the eight bivalents at diakinesis and metaphase I. Some PMCs have only eight bivalents. The number of B-chromosome in other PMCs ranges. In all the cells observed, one bivalents is exceptionally large. The segregation of chromosomes, including the larger one, is perfectly regular at anaphases. Whereas the distribution of B-chromosomes is random with incidence of lagging being quite high, pollen fertility is almost 100%. Amongst the agricultural Nepalese populations, cattle breeding is of great importance even though the returns are often quite low.

Right across the country, from north to south, animal husbandry is practised in its various forms, adapted to the widely differing natural environments and deeply marked by the cultural heritage of the different groups that have taken to this activity. For reasons that are both historical and cultural, resources are raised in a number of ways depending on the requirements of the ethnic groups. The Indo-Nepalese populations are principally rice growers. For them, owing buffaloes, zebus, goats, sheep is a marginal economic activity, which is none the less indispensable as animal waste is their only source of field indispensable as animal waste is their only source of field manure. Even though every family tries to have a cow-an animal venerated by the Hindu religion-under their roof, the number of animals per household is low and there are hardly any large herds.

The herds never graze on the upper portion of the high mountains (altitudes higher than 3300 m) and in the course of their daily or annual transfers never go beyond the upper limit of the forest. As they move within a limited space, they do not require a specialised habitat. On the contrary, cattle breeding is an important activity for the Tamangs, Magars and Gurungs who have settled on the southern slopes of the high Himalayan chain, where maize and eleusine are grown. Apart from the buffaloes and zebus, from which the family gets it milk supply and which are needed in the fields, transhumant breeding of sheep and goats is practised on a larger scale.

Sheep are raised primarily for wool, without which clothes cannot be stitched, whereas goats provide meat, consumed especially at the time of major festivals which heighten the life of the village communities.

The herds move over a wide area (altitudinal and geographic), summering in high altitude pastures (bet, 4000m and 5000m), sometimes at a great distance from the village of origin, and spending the winter months in the cultivated zone of village lying in the middle valleys (bet, 1000m and 1500m). For the populations of Tibetan origin, Sherpas included, settled in the north of the country, whose small agricultural holdings are situated in the upper lands suitable for cultivation (above 2200m), cattle breeding is a traditionally favoured activity. The yak hybrids (cross between a yak and a zebu) allow one to make more extensive use of the different ecological stages than pure yaks which cannot adapt easily to "lower" altitudes (less than 3000m) and are thus confined to the highest zones. Such transhumant breeding is intended for the supply of dairy produce as the milk obtained is converted entirely into butter or dry cheese, either for self-consumption or sale at the market place.

In summer, the herds are taken to pastures situated either on the southern slopes of the high chain, or to the north in the interior valleys beyond the Chinese border on the Tibetan plateau. The animals spend the winter in the lower forest regions (mountain side oak groves) or in villages where winter fodder reserves are used to feed them. The different forms of transhumant breeding have led the Nepalese sheperds to adopt constructions which are well adapted to the environmental constraints and meet the technical imperatives specific to the genetic type while lending in with the traditional model of village construction. Settlement, orientation morphology, materials used, organisation of internal space: it is these characteristics, specific to such a construction, which represent the optimal solution of the breeders in a given environmental and cultural context.

All settlements require a watering place nearby and a steady supply of firewood. Water and wood, which are indispensable in the domestic life of the herdsmen, are required in larger quantities for breeding hybrid or pure yaks. While sheep and goats need only a limited amount of water, large animals require a lot of water everyday; besides, the techniques used for producing butter and cheese from the milk of the hybrids (or yaks) require large amounts of water and wood. Some high mountain pastures, either too far away from the upper limit of the forest or not having an adequate supply of water are consequently visited only by sheep and goats.

During the monsoon however, water is rarely a problem: in the absence of a watering place (source or waterway) in the immediate environment of the grazing space, a hole is made in the ground to collect rain water. Sometimes, bamboo drainpipes running along the edge of the roof of the shelter enable water to be collected in a large wooden barrel. Shelters are therefore located preferably near the forest, close to a watering place. As far as possible, shelters are constructed on sites which are relatively flat, away from very humid locations (according to the shepherds, goats feat humidity the most). The raised course of the shelter provides efficient insulation from soil humidity. Drains or channels dug around the shelter facilitate drainage.

In the high mountain pastures, the shelters of hybrid breeders are often situated on a small open courtyard covered with slabs of shale. It is here that the animals spend the night and are milked in the morning and evening. In the forest, the pastoralists build their shelter in the glades. The shelter is sometimes surrounded by an enclosure made of rhododendron branches to prevent the goat and sheep from moving out at night and becoming easy prey to bears and panthers. The location does not change from year to year. The orientation is variable. The longitudinal axis of the shelter is always perpendicular to the line of the slope and the only opening, used as an entrance, more often than not looks downwards.

Typology, morphology and material is to be recalled that this text is dealing with the habitat of transhumant pastoralists. As grazing pastures are often small in area and relatively dispersed, the herds have to move quite frequently in the course of the annual pastoral cycle. There are two main categories of constructions: mobile shelters used all year round and fixed shelters used on a temporary basis. The mobile shelters have a semi-cylindrical shape. The wooden framework, which is nothing more than a ridge pole and a few stakes driven in the ground, supports the roofing made of bamboo mats, 'bakhari', held in position with flat heavy stones at ground level.

The hermetic semi-cylindrical shape (a chick constitutes the only opening and this is pulled down at night or when it rains) makes the construction resistant to wind and provides effective protection against the cold. The vertical inner walls are of basketwork, of rhododendron or juniper branches, or of rough stonework. When the herd moves towards another pasture, the shelter is dismantled; the basketwork is rolled up and transported along with a part of the wooden infrastructure to the new place of pasture where it is reassembled.

The dimensions of the bamboo mats used hardly vary from region to the shelters are more or less the same throughout the country. On the contrary, the length is extremely variable; it ranges from a few metres (5m on an average) for the shelters of the shepherds to 25 metres for the shelters of yak or hybrid breeders (the transformation of milk into butter and cheese and the need to shelter the calves at night require a larger habitability). A second type of mobile shelter can be found only north of the high chain, amongst the breeders of Tibetan language and culture, practising transhumant breeding. The tent, traditional dwelling of the Tibetan Drog-pa nomads, is their customary shelter. Bands woven from yak overhair and stitched together constitute the canvas supported by two poles and held in place by runners also made of yak overhair. Unlike the mobile shelters, the fixed shelters, 'ghiang', are used on a temporary basis.

A double sloping truss constitutes the roof. The two sides and the ridge pole, wedged into the walls, rest on two vertical beams. Depending on the material available from the immediate environment, the walls may be of raw stones obtained from glacial fallen rocks, pine shingle or even basketwork. The roofing could also be of single or roofing stone, simply placed on the frame. At the beginning of autumn, when the herders leave the high pastures for the forest below, the shingle roofs are dismantled to prevent them from crumbling under snow; the shingles are stored nearby. Such construction, more vast than the preceding shelters, are generally colder.

Apart from the Drog-pa tents, there is very little variation in the internal organisation of the different types of shelters. The entrance, which more often than not constitutes the only opening, gives on to grazing zone. The interior is divided into two rooms; one enters directly into the inhabited room; the ground is covered with a pine floor that has not been fixed (fixed shelter) or juniper branches. The inhabitants of the shelter spend the night near the fireplace dug in the floor, lying on bamboo mats or animal skins (calf or sheep skins); wood is stored in a drier above the fireplace. The inhabited space is also a place of work: the churn and milking buckets are kept here; the different containers and utensils are arranged on the recesses in the wall (fixed shelters) or placed near the fireplace on a shelf at ground level. The calves (yak hybrid breeding) or the lambs and goats (ovine/caprine breeding), spend the night tied to a rack in the second room. The ground is covered with a little of pine or juniper branches changed regularly.

The shepherds and cowherds spend their day tending the herds, milking the animals, manufacturing milk products as well as performing various domestic tasks. For them, far removed as they are from their village, isolated in the high mountain pastures flooded with monsoon rain, the shelter constitutes an indispensable fixed point around which their activities revolve, a place where, when evening falls, they find a little comfort and warmth. We will describe a day in the month of August at Dumkeur (4030m), high pasture situated on the southern slopes south of the Himalayas, north-east of Kathmandu, spent with Sonam's herds.

Ributi Sherpini, the wife of Sonam from Khartal, is living in the high mountain pastures. There she will stay for the whole year with her herds of goats and hybrid yaks. She hardly ever stays in the village. She is 45 years old and lives with her two daughters, 12 and 8 years old respectively, and her two sons, one 10 years old and the other six months. It is 5 o'clock in the morning. Outside, morning has almost broken.

A huge cloud formation settles in a corner of the valley, several kilometres below, covering the cornfields which surround the village, situated at a day's trek. The dog frisks around the calves, at the spot where he spent the night. Ributi, probably awakened by the morning chill (5°C in the shelter), emerges from her woolen blanket, fully dressed. Her first concern is to rekindle the fire. Sitting cross-legged near the hearth, she uses a bamboo blow torch on the glowing embers which can be seen through the scattered ash.

The dry sticks in the hearth immediately catch fire and start cracking. The little girls and the elder son, Nim Sona, disturbed by the noise, toss about under the blankets and finally get up. The baby continues to sleep in the bamboo cradle. Morning has broken (5.30 a.m.). The temperature begins to rise. The mist of clouds lifts slowly and covers the nearby pine forests. The hybrid herd is dispersed around the shelter, some animals are given the driest spots to sleep and only the calves sleep inside the shelter. Conscious of the proximity of their mother, they pull and tug at the rope attaching them to the rack. The eldest daughter takes the plastic jerry can and goes to fetch water from the source, situated on the ground at a distance of a five-minute walk from the shelter. The younger daughter cradles the baby who gas started to cry. Nim Sona and his sister raise the barrier at the entry of the goat shelter, letting them out whereas the young ones remain inside.

The mother cleans the area surrounding the shelter; the dung collected by hand is piled up near 'Berberis' bushes; the children help their mother and then leave to herd the hybrids. It is 8o'clock, time for milking the animals in the morning. Ributi washes her hands, ties a small leather bag containing salt around her neck, takes a rope woven from yak overhair, picks up the milking bucket and calls the animals outside the shelter. nim Sona unties one of the calves which immediately rushes towards its mother and starts suckling. Ributi allows it do so far a minute (just about 30 seconds), then forcibly pulls it away by its rope collar and attaches it to a stake near the entrance of the shelter. She gives a handful of salt to the mother who begins to lick her calf. With the help of a rope, she immobilises the hind legs of the animal, then squats, the wooden bucket placed between her knees and starts to milk, her head resting against the side of the animal. The milking over, she unties the mother, makes it move away and then releases the calf which laps up whatever little is left of its mother's milk. Ributi then goes into the shelter and pours the contents of her bucket in one of the huge copper cauldrons that has just been washed by one of her daughters. She goes out again and calls another females animal: "Naptema, che che !". The animal responds to the call and approaches, bleating. Nim Sona unties the second calf. Milking is carried out in the humid mist which will persist for the rest of the day. The order in which the animals are milked does not change from day to day and some animals seem of understand this and come automatically when their turn comes. All the while the two little girls milk the goats. One of them holds the animal by the head to keep it still, while the other, in a squatting position, milks the animal between its hind legs in order to collect the small quantities of milk in a white metal can.

It is 8.45 a.m. and the milking is over. The younger daughter adds wood to the fire as Ributi and her elder daughter place the heavy cauldron on the trivet of the fireplace after having added a few ladles of water (about a litre) to the milk. While the milk is being heated on a slow fire, the mother breastfeeds her baby. As soon as the skin forms on the surface of the milk (some twenty minutes later), the cauldron is removed from the fire. In order to facilitate fermentation, Ributi adds a ladle of buttermilk of the previous day. Once covered, the cauldron is put at the back of the inhabited space.

Ributi then prepares the morning meal while the children bring back the calves one after the other and tie them to the rack. The elder daughter goes out to fetch water, the younger one cradles the baby as the cornflour gets cooked on the stove. Ributi grinds salt and pepper in a mortar; the meal is ready and the mother serves her children. The meal is taken sitting cross-legged near the fire place, the cooked cornflour is eaten not with a spoon but with the right hand, out of an aluminum bowl. A soup spoonful of curdled milk completes the meal; each one then washes his plate and puts it on the shelf near the fireplace. Ributi takes the baby who has again started to cry.

After breastfeeding the child, she lays him down in the cradle and covers him with a blanket, after which she goes out to collect her herd which she will guide to the forest. The animals are guided with the help of different kinds of missiles (stones or lumps of mud) thrown in the direction of those that linger on. The herd of goats, shepherded by the little girls, has to move in the direction of the steep cliffs, inaccessible to the hybrids. The elder daughter spends the day with the herd of goats and will return in the evening with a basketful of hay or bamboo stalks for the calves.

The hybrids are left untended; Ributi comes back to the shelter, takes her 'khukri' knife and goes to cut wood with her son; the baby is still sleeping near the fireplace. It is 12 o'clock. Ributi and Nim Sona have come back, each one loaded with a heavy burden of moist wood; the wood is stored in the drier. Nim Sona unties the calves, which browse, dispersed around the shelter.

The younger daughter has come back. She deposits her basket of fresh grass near the litter of the calves, places some of it on the rack and stores the rest so that it is out of reach. All this while Ributi transfers into the churn that had been washed earlier the milk obtained the day before (in the evening) which has curdled. She then pours cauldrons and throws the rinsed water in the churn. Taking the beater kept at the joint of the roof, she fits it on the supports and starts churning. The younger daughter returns with water which she puts to heat in the cauldrons and then stokes the fire one again. Ributi continues to churn at a sustained pace, stopping occasionally to recover her breath (the beater is normally turned thirty times per minute). The first grains of butter appear on the surface. The elder daughter then pours in from time to time ladles of hot water (60° appx) on the inner wall of the churn and on the lower support of the beater. It is 2.30 in the afternoon. Ributi removes the beater (churning has gone on for 45 minutes) and then washes her hands. She collects the grains of butter which are floating on the surface and deposits the butter on a large wooden ladle. The lump of butter thus formed is put in a small pan filled with cold water. Once the beater has been cleaned, it is put back in place under the roof. Ributi collects the grains of butter still floating on the surface of the buttermilk and uses them to oil her hair.

A part of the buttermilk is kept aside for the evening meal; the rest is emptied into the copper cauldron and put to heat on the fire. The younger daughter once agains adds wood to the fire. The butter is stored in a wooden box kept in a corner of the inhabited space. The buttermilk proteins begin to coagulate on the fire (after 20 minutes of heating). Ributi slowly lowers a wicker basket into the contents of the cauldron; the whey that penetrates in it is removed with a ladle into a pan and then kept in a wooden barrel. This operation is repeated till only the coagulate is obtained which is drained in a basket and stored in a wooden pail.

Meanwhile, Nim Sona cleans the litter of the calves, arranges the pine branches on the ground and spreads the fresh hay brought from the pastures. Ributi continues making cheese: she deposits the coagulate in a square shaped canvas that she had dried in the morning. After having joined together the angles of the canvas, she squeezes it and makes a knot. She then places the canvas with cheese outside, pressed between two large flat stones. The next day she will remove the cheese from the canvas, cut it into small cubes and put them to dry in a basket hanging above the fireplace for this purpose. It is 3.30 p.m. The younger daughter and Nim Sona who have gone to search for the calves will bring them into the shelter. They will distribute to them the whey that is till warm before tying them to the rack. Ributi will feed the dog a mixture of whey and cornflour placed in a small wooden trough. At around 5o'clock in the evening, it begins to rain.

The elder daughter returns with the herd of goats and the hybrids return on their own from the grazing lands. Nim Sona and the younger daughter leave in search of those who left behind. Ributi is in the meanwhile distributing whey to the females that were the first to come in, and the elder daughter releases the kids which had spent the day in the shelter. It is 5.30 in the evening. Time for the evening milking in the rain. As in the morning, Nim unites the calves, one by one, but this time he will only tie them back after their mothers have been milked.

Milk is put to heat. Outside, night has fallen (6.30 a.m.). Ributi removes the milk from the fire and places the cauldron in a corner after having added a little buttermilk and covered the vessel with a lid. Cornflour with a little curdled milk constitutes the evening meal. The day is over. The family gathers around the hearth and spend the evening watching the fire die out slowly.

The day's activities take place between the rising and the setting of the sun. During the monsoon, the sun rises at 6 o'clock and sets at 6.30 in the evening. The sunshine is dependent on the cloud covering which in this season is permanent; during this period, the inside of the shelter is plunged in darkness for the most part. Pastoral activities set the rhythm of the day. In the evening, the hybrid herd often assembles on its own around the shelter, reminding the inhabitants with clockwork precision that the time for milking has come. Most of the people living in the high mountain pastures do not have a watch but tell the time by looking at the position of the sun, when it shines. It must be noted that the daily occupations of the shepherds are different from those of other herdsmen and this for two reasons: contrary to the yaks and their hybrids, goats and sheep require to be watched over continuously; moreover, the milking of goats and sheep is only a subsidiary activity and milk technology only has a small role to play in the shepherd's schedule.

3 Forest Ecosystems

The ecosystem-level studies are based on a series of representative forests which cut across all the three vegetational regimes. The rainfall pattern is governed by the summer monsoon. The year has a warm dry period, a warm wet period, and a cool dry period. The climatic mesicness tends to increase with increasing elevation. The annual rainfall varies between about 1300mm and 2500mm. As expected, temperature decreases with the rise in elevation. Snowfall occurs above 1800m. Since the region lies within the subtropical latitudes (28°44'-30°44' N), the day length variations within the annual cycle are limited.

Towards the higher elevations temperatures approach those encountered in the temperate regions. Under sal forests, located in "bhabar", 20-25km wide talus gravel slopes, deposited by the rivers, the soil is relatively deep, and alluvial, while under the forests of higher elevations it is residual and shallow. Towards the lower elevation it is loamy sand, while towards the higher elevations it becomes sandy loam.

The bulk density of the stone-free soil declines with the increasing elevation and the organic matter and N contents increase. The pine forest is distinguishable from other forests in having large proportions of stones and markedly higher C:N ratio of the soil. Analysis of a collection of 1200 plants species from the Nainital region indicates that 426 species are used variously as medicinal, oil, fibre, timber and fodder plants. These species belong to various growth-forms, such as trees, shrubs, herbs, climbers, parasites, and epiphytes. They may cover a wide range of-habitats and ecosystems, such as exposed and dry rocks, moist and shady places, stagnant water bodies, streams and wetlands, croplands, grasslands, rocks, man-made walls, tree trunks and branches. Because of the degeneration of habitats about 85% of these species (mostly herbs and shrubs) have become scarce, many need conservation and some delicate herbs are on the verge of local extinction. The Corbett national park, located in the Naini Tal District, alone, harbours 46 mammals, 94 birds, 11 reptiles and 47 fishes of great interest.

A rich wildlife is found in the native forests, but with the increased destruction of forests more and more species are becoming extinct locally. For example, the red-giant flying squirrel, whose habitat is the oak forest, is becoming increasingly rare in several areas where intact oak forests are no more found. Tewari and Pant suggested that the bryophytes are a promising biological tool as indicators of mineral deposits. About 14 liverworts and 48 mosses can be used to locate the mineral-rich substrates with varying levels of accuracy.

In general, the species richness and diversity of these forests are comparable to those of the temperate forests. However, some transitional communities, such as chir pine, pine-mixed broadleaf and sal-pine-tun (Toona ciliata), which lie between sal and pine regimes, approach the diversity levels reported for tropical forests. In a stand of pine-mixed broadleaf forest, located between 1200-1500m elevation, 27 tree species per ha were reported by Tewari. The dominance-diversity curve for this forest assumes a lognormal nature, attributed to species-rich forests of favourable environments. In contrast, in the pine forest where single species-dominance is strong, the dominance-diversity curve follows a geometric series attributed to species poor communities of unfavourable environment. Sal forest and oak forests exhibit intermediate situations. Our studies indicate that within 1200-2000m, pine forest represents an early successional stage, pine-mixed broadleaf forest an intermediate stage, and banj oak (Quercus leucotrichophora) forest the climax stage.

Bryophytes associated with mineral enrichment in Kumaun Himalaya

Minerals	Associated liverworts and mosses
Mica schist	Grimmia pulvinata
Granite Rhacomitrium	Grimmia pulvinata, Grimmia sp., himalayanum
Iron oxide splachnobryoides	Entodon luridus, Fissidens
Iron hematite luridus,	Plagiochasma appendiculatum, Entodon Fissidens sp, Hyophila involuta, Macrothamniella pilusula, Oxystegus sp,
Tortella sp.	
Iron limonite	Jungermannia duthiana, Bryum sp, Campylopus flexuosus, Desmatodon sp, Dicranum sp, Drepanocladus uncinatus, Fissidens sp, pogonatum sp, Pohlia crudoides,Rhacomitrium sp, Scopelophila ligulata.
Graphite schist	Trichocolea tomentella, Bryum cellulare.
Gypsum	Asterella sp, Cephaloziella sp, Riccardia multifida, Anoectangium stracheyanum, Barbula horricomis, brachythecium buchanani, Bryum bessonii, B. capillare, Campylopus sp, Dicranella sp, Fissidens sylvaticus, Gymnostomum calcareum, Philonotis mollis.

Calcareous tufa	Dumortiera hirsuta, Pellia endivlaefolia, Chiloscyphys polyanthus, Plagiochila sp, Atrichum obtusulum, Barbula gracilenta, Fissidens grandifrons, F. taxifolius, Eurhynchium ripariodes, Hydrogonium javanicum, Philonotis catearea Vesicularia montagnei. (Not forming 'fossil casts')
Asterella sp,	Marchantia palmata, M.polymorpha, Plagiochasma appendiculatum, Jungermannia cordifolia, Cratoneuron commutatum, Mnium cuspidatum.
Magnesite	Hymenostylium recurvirostrum var. cylindricum.
Soapstone	Marchantia polymorpha, Frullania sp, Anoectangium sp. Desmatodon sp, Philonotis speciosa, Pohlia camptotrachela, Splachnobryum sp.,
Dolomite	Anoectangium thomsonii, Hymenostylium recurvirostrum, Hyophila involuta, Tortella tortuosa, Zygodon viridissimus.
	Copper, lead and zinc Marchantia sp; Barbula sp. Mielichhoferia sp.

The tree basal cover of some of the stands of these forests (60-80 m^2 ha^{-1}) fall towards the higher side of the normal range given for the forests of the world. The biomass values for these forests also indicate that the entire elevational range has potentiality of supporting large biomass. Additional biomass ranges are given in Negi et.al, and Tiwari and Singh.

The shrubs attain the maximum expression in oak forest followed by sal forest and are least developed in pine forest. Quantitative profile structure of several forests was reported by Saxena and Singh. As indicated by the profile diagram, the oak forest is multistratal, in contrast to the simpler pine forest. Such a multistratal forest dampens the thrust of rain drops, as a consequence of which soil and nutrient losses are minimised.

Total basal area and biomass of the Kumaun Himalayan forests located between 300-2200 m elevation

Forest	*Total basal cover* ($m^2 ha^{-1}$)	*Total biomass* ($t\ ha^{-1}$)	*Total litter fall* ($t\ ha^{-1}$)
Sal oldgrowth	56.83	716.9 (3223.7)	7.5
Sal seedling-coppice	39.93	460.4 (286.0)	6.7
Pine mixed broadleaf	24.25	199.0 (215.9)	4.1
Pine	35.92	200.8 (375.2)	7.6
Mixed oak-pine	40.48	432.6 (156.9)	6.3
Mixed oak-Rianj dominated	59.68	562.2 (262.0)	7.4
Mixed oak-Tilong dominated	72.08	786.7 (200.9)	7.1

During the rainy season, within 1500-2050 m elevation, the epiphytic flora in several forests of Kumaun Himalaya contribute to the total community chlorophyll, amounts greater or equal to the combined contribution of herb and shrub layers (Singh and Chaturvedi, 1982) However, the per cent contribution to the total amount of community chlorophyll by epiphytic flora varies considerably from forest to forest, being about 16%, each, in banj oak (Quercus leucotrichophora) and tilonj oak (Q. floribunda) forests, about 14.5% in surai cypress (Cupressus torulosa) forest, and only about 3% in chir pine forest.

Lack of prolonged seed dormancy in a number of species, and the inability of soil to store the seeds in a viable state for long periods of time, may lead to disappearance of these species in the face of persistent disturbance. Wide response breadths in terms of seed germination on the temperature and light gradients (average levins' response breadth B for most of the 25 native species studied was 0.929 on temperature gradient and 0.929 on light gradient), and relatively narrower response breadths on the moisture gradient (average B = 0.769) suggest that the distributional range of most species is restricted by moisture.

Investigations on seedling growth along a light gradient indicated that both the banj oak and the pine exhibit maximum growth in 70% of the full sun light, and the former showed a wider response breath than the latter. Evidently, availability of more light on the forest floor, as a result of the disturbance of the forest canopy, is not likely to favour the regeneration of pine against that of banj oak. However, pine may be have competitive advantage over banj oak in situations of hydric deficiency.

No conclusive results were obtained in regard to the regeneration of sal, the most important timber species of the region. It seems that the successful regeneration of this species largely depends on its ability to attain the sapling size class (individuals with cbh between 10-30cm), before which the individuals survive the unfavourable environmental conditions (e.g., hydric deficiency, frosts, etc,) through repeated die-backs. If the die-backs were the inherent property of the species under normal conditions, it may be suggested that sooner or later a phase of good regeneration, i.e., when individuals are able to reach the old sapling size frequently, will occur, and the "phasic regeneration can be regarded as a property of the species. The gap-phase-type of population structure suggests such a trend. However, if it was a site specific character, the environmental factors that influence it, need to be investigated. Over large areas, oaks are failing to regenerate and successional species, such as pine are encroaching upon these forests. Several factors have been implicated in the poor regeneration of Quercus leucotrichophora, the most important ones being lopping (reduction of seed crop). Seed predation, and grazing.

Interpretations based on density-diameter distribution of trees individuals with cbh 10 cm and above) over about 8000 ha area indicates large scale replacement of oldgrowth forests by more even-aged, successional forests. Density-diameter curve (d-d curve) for the entire assumes an overall convex appearance (largely, second derivative negative). It is likely to be constituted as a result of exploitation practices, which maximise the even-agedness and the proportion of intermediate-age stands.

The relatively shade-tolerant oak species are failing to regenerate and shade-intolerant species, such as pine, are regenerating abundantly in younger forest stands where canopy shade is light. The overall total forest structure thus gives an impression of an early successional forest. The exploitive management practices, and the biotic stress exerted by the hill population (grazing, lopping, felling for firewood) on oak forests. Even in pine and sal forests, population of intermediate ages appears to be promoted. The oldgrowth, all-aged multistratal forests are being removed at a faster rate than they are being reconstituted. The exploitation cycles appear to be shorter than they should be to promote unevenaged forests.

A principal hydrological character of the Central Himalayan catchments is that these are sub-surface flow systems, the overland flow accounting for only a small portion of the incident rainfall (annually less than 1%).

The majority of the rain-water is transmitted laterally to the channels via a sub-surface "quick flow" process which can produce typical storm hydrographs in the main streams. Such sub-surface flow systems with shallow soil are usually vulnerable to landships and lanslides, as the soil saturates rapidly.

A study of the recovery of oak forest ecosystem damaged by landslide indicated that the ecosystem may return to the original stage quite rapidly, provided that the site is not persistently stressed biotically. The climate is benign, and the favourable conditions during a major portion of the year for woody growth enable the rapid reconstitution of the original tree composition. The concentration of nutrients in the soil as well as the fine-soil content increased with time and returned to the characteristic level of the reference forest stand (adjacent undisturbed forest) in a period of about 40 years. Generation of overland flow decreased with the development of the ecosystem and achieved an equilibrium when the herb and the shrub layer vegetation had fully developed.

The nutrient loss through runoff water and soil decreased, and the nutrient cycling became tighter with the development of vegetation. Seedlings of Quercus leucotrichophora and Q. floribunda appeared after 21 years of the landslide. Thus, although some degree of improvement seems to be prerequisite for the recruitment of these climax species, they can establish directly without following a sequence of shade-intolerant and intermediate species. It should be possible to shorten the time lag in the appearance of climax oaks through suitable management practices such as dibbling of seeds and establishing a proper herbage cover.

We examined the seasonal periodicity of a number of activities, viz., leaf drop, leafing, flowering and fruiting of trees and shrubs, productivity and litter decomposition in the forest ecosystems. Similarity rather than dissimilarity among the forest ecosystems is the striking feature. The monsoon pattern of rainfall seems to be the most dominant common feature among the forest ecosystems located between about 300 to 2200 m.

Most species tend to derive the maximum advantage of the rainy season, the forests of this region resemble one another considerably in their phenology. In all the forests the evergreen woody species with concentrated summer leaf drop prepondered. (Shorea robusta, the dominant species of the sal forest, is considered deciduous, though its trees are seldom naked). Both the leafing and the leaf drop occurred simultaneously and the former seems to induce the latter. The leaf expansion is completed with the advent of rains, and the canopy is maximally developed during the rainy season.

Fruiting periodicity is such that in a large number of species, seed germination synchronizes with rainy season. However, in another group of species, the percentage of which is greater in the forests of higher elevations, the fruiting periodicity is such that the seed germination occurs during the extra-rainy season. In this way, seeds and fruit are available almost throughout the year for the wildlife. More than one "guild" of herbs are encountered in each forest. They differ from one another in the time of peak photosynthetic activity. In general, the periodicity of production in herbs differs from that of the woody species.

The two woody forms (trees and shrubs) also show subtle differences in their periodicity of production, the shrubs tend to attain the peak earlier than the trees. These differences minimize the competition among the various growth for the resources, and maximise the overall efficiency of the ecosystem. In general, the rate of decomposition is maximum during the rainy season. Between the two groups of decomposers studied, microarthropods assume dominance earlier than the microbes (fungi, bacteria and actinomycetes).

The net primary productivity (in terms of dry matter) in Kumaun Himalayan forests ranges from 12.6 to 27.6 t $ha^{-1}yr^{-1}$. There is a considerable variation in the net production from forest to forest, and even fro one plot to the other within the same forest, depending upon the size of the tree population. Litter fall was studied by Pandey and Singh, Mehra and Singh and Singh, Rawat and Chaturvedi.

Pandey and Singh, Tewary et al, and Upadhyaya and Singh have studied the litter decomposition in these forests. The rates of decomposition of leaf litter of certain broadleaf species of sal and pine-mixed broadleaf forests (0.253% per day) are comparable to those given for tropical lower montane forests. The rate of decomposition of pine leaf litter is considerably lower than for other species. Markedly high C.N ratio and lignin: nitrogen ratio (35 in pine compared to 15 in oak litter), and low microarthropod and mocrobial populations are the major factors responsible for extremely low rate of decomposition of the pine litter.

The rate of decomposition was positively related to the temporal beta diversity of the fungal communities. The fungal community developed on the slow-decomposing pine leaf litter was conspicuously different from those developed on other litter species. In several species the nutrient release from decomposing litter was preceded by nutrient, particularly N, immobilization. In sal forest, termites seem to play an important role in nutrient cycling. Their respiratory and net secondary production energy cost accounted for about 1% of the annual energy input through litter fall.

The climax forest communities (sal forest, oak forests) resemble more the tropical rain forests than the forests of the temperature zone, from the stand point of nutrient allocation in different components of the ecosystem. Compared to the temperate forest ecosystems, a greater proportion of nutrient mass is stored in biomass in these forests. For example, of the total accumulation of N in the ecosystem in mixed oak, tilonj-dominated forest, nearly half is accumulated in biomass; the same is about 70% for P, 80% for Na and above 90% for K. Comparatively higher concentration of nutrient in wood of these forests than that of the temperate forest is largely responsible for this situation.

Rate of leaf and wood litter decomposition for certain species of Kumaun Himalayan forests

Forest Species	*Decomposition rate (% d^{-1})* Leaf	Wood (4-6 cm diameter)
Sal Shorea robusta	0.253	0.122
Pine-mixed broadleaf Quercus leucotrichophora	0.253-	
Lyonia ovalifolia	0.253	0.067
Pine Pinus roxburghii	0.126	0.060
Mixed oak-pine		
Quercus leucotrichophora	0.165	0.084
Mixed oak Quercus lanuginosa	0.150	0.067
Quercus floribunda	0.150	-

Inputs of nutrients to the soil in these forests have been quantified by Pandey and Singh. Input of N is predominantly through leaf fall in all ecosystems.

A substantial portion of the input of K (33-46% of the total input), P(23-29%) and Ca (18-26%) is through the precipitation components. Singh, Rawat and Chaturvedi illustrated how the nutrient cycling strategy of pine contrasts with that of oak. Compared to pine, the oak soils is eutrophic. The C:N ratio of oak soil was 8.5 compared to 14 in the pine soil. Similarly, the C:N ratio of accumulation in net primary production was significantly higher in pine. While substantial quantities of nutrients were reabsorbed from senescing leaves before abscission in pine (about 50% each of N and P of the leaf content), it was quite low in oak (about 13% n and 8.5% P). The efficiency of litter production per unit of N in pine far exceeds that of oak. Thus, the nutrient conserving efficiency is much more developed in pine than in oak.

Both settled agriculture and pastoralism cause massive forest exploitation. Tea gardens, mining, quarrying and construction of road-network have caused a substantial loss of forest cover. Starting from about 1860s to 1870s the impact of commercial exploitation, first for railways, then to meet the demands of the world wars and subsequently of urban expansion, etc., has been considerable. These aspects have been discussed in detail in other contributions to this volume. Here we focus on the functioning of the settled agriculture, on which most of the population in mid-Himalaya depends. The strategy for the environmental regeneration would depend largely on our ability to develop a viable form of cultivation which is in harmony with the environment.

Before analysing the ecology of agriculture it would be useful to throw some light on the current rural life in the hills: (i) The population density in the most habitable elevations (1000-2000m) generally ranges between 100-120 per km^2. The population is increasing at an exponential rate of 2.28% per year and the agricultural land at 1.5% per year. (ii) Generally, a house hold consists of 5 or more members: the cultivable land per household for a great majority, varies from 0.45 to 1.05 ha: and most (80-90%) of the cultivation is rainfed. (iii) The increase in the male emigration with time has increased the proportion of females in the resident population and has led to and increase in the participation of women in various primary activities including agriculture. (iv) More 'development' also means decrease in forest resources available to villagers, which in turn leads to greater expenditure of human energy for fodder and firewood collection. (v) Emigration of younger working age-group increases the proportion of older population in the villages, which is inefficient in adopting the facilities provided for development. (vi) The human diet is essentially vegetarian and the input of food from outside the hills is considerable. (vii) Despite the decrease in the availability of forest resources, the animal population is not declining, thus the pressure on the land for fodder is increasing far beyond the carrying capacity.

The settled agriculture

Pandey and Singh and Singh, Pandey and Tiwari have recently analysed the energies of the settled agriculture of Kumaun Himalaya on the basis of sample villages. In these studies a complete inventory was made of all inputs (labour in terms of man and animal days, fertilizers in terms of manures and chemicals, and seeds), and outputs (yields of edible crop products, crop by-products, and milk).

The sources and supply of animal fodder and the use of firewood by the humans were also quantified. It became evident that: (i) the agronomic production largely depends on the energy supply from the adjacent forest ecosystems, and (ii) the amount of edible products from the agriculture is far less than that is required to meet the demand of the village population. Each unit of agronomic production is obtained at the expense of about 7 units of energy from the adjacent forests in terms of fodder and firewood, and leaves for manure. Against the requirements of 18 ha of forest land (including 5-12 ha of well-stocked forest) per ha of cultivated land, at the present level of agro-activity, the ratio of agricultural land to forest is only 1:1.33 and the ratio of agricultural land to well-stocked forest is only 1:0.84. Thus, the carrying capacity of the forest has already been far exceeded. The protein-rich broadleaf tree species such as, Quercus spp., (oaks), Bauhinia spp., Ficus spp., Mallotus Philippines, etc., have been the worst suffers, as they are frequently lopped for leaf fodder and fuel wood. If the present form and size of agriculture continues, dung will be increasingly used for fuel, and consequently the agricultural production will sharply decline.

The present forest-based annual crop cultivation is already on the verge of collapse in some areas. Further, we need forests for the very survival of the Mountain ecosystems. The forest cover is already reduced to a small fraction of what is required for the ecological protection of the hills. The terrain, the size of land holdings, and irrigation facilities are such that the mechanized fossil fuel-based agriculture cannot replace the present form of agriculture. Farm forestry can be one suitable alternative. The first prerequisite of this model is that a village will raise fodder and firewood species on his own land, so that his dependence on the state-owned forests is minimized. However, where irrigation is available he will also raise fast-growing tree species of commerce.

The relevant developmental agencies, such as, the-forest department. Universities and other institutions will help him in selecting and raising the suitable species, which will vary in accordance to the environmental factors, Emphasis should be on raising multistratal, mixed plantations so that all possible ecological niches are filled and overall efficiency of the tree farm system is optimized. However, since in this model it is suggested that the villagers will not derive energy from the reserve forests, the effective area for his use is considerably reduced. Therefore, a much larger size of land holdings will be necessary. Radical administrative means would be required to transfer more land to the villagers for the sole purpose of growing trees. Apart from reducing the pressure on the reserved forests, this kind of farm forestry would have the following advantages (i) The disturbance of soil involved in perennial tree plantation would be negligible compared to that in raising annual crops, as repeated ploughing and other soil treatments will not be required. (ii) The multispecies perennial tree plantations, by utilizing the solar radiation year round, are likely to be more productive than the present agriculture in which, on average, two short-lived annual crops are grown on a piece of land in three years. (iii) The livestock population would be reduced to a manageable level for ploughing, and organic manure will not more be required. (v) Presently, 2.5 times more energy is spent in the collection of fuel and fodder than that spent in direct agroactivity.

A large proportion of this can be devoted to other village-based and forest based economic activities as discussed by Singh. Pandey and Tiwari and in several papers in Singh et al. The pressure of work on the women would be eased. (v) The farm forestry, by facilitating the development of forest cover on the mountain surfaces, will ensure the reduction in sediment load of rivers. The villagers will also be expected to develop pastures on lands unsuitable for farm forestry. However, the species will have to be screended for tolerance to water stress and for nutrient value. It is further assumed that the granaries of the plains will supplement the output of the grain in the hills.

In other words the tree plantations in mountains, in part, would be raised at the expense of fossil fuel-based agriculture of the plains. The mountains in turn, will protect the plains against floods, in addition to providing timber, paper pulp and other commercial forest products. In the Indian central Himalaya, as pointed out earlier, the area under forest is far less than generally assumed. Forest ecosystems are needed to protect wildlife (including medicinal plants) and soil, to sustain man, to stabilize climate, to optimize water yield and to purify water. Although the region has the potential of supporting large forest biomass over extensive areas, only a small fraction of the potential biomass exists.

Early successionals predominate the forests, and old-growth forests are scarce. Unregulated and excessive exploitation of forests and faulty management practices seem to be largely responsible for this situation. The felling cycles should have been longer than that were followed. If the same management practices continue, more and more original forest communities will be converted into simpler and degraded vegetation. Inherent functional properties of species have to be taken into consideration. For example, oak has a relatively heavy demand for N. Through the return of N-rich litter and rapid mineralization, oaks maintain high soil fertility. Heavy lopping of oaks cause large canopy gaps and reduce leaf fall and hence N return, making the conditions suitable for invasion by pine.

The recalcitrant litter of high C:N ratio, greater capacity of litter production per unit of N efficient N retranslocation before sensecence, microbial immobilization of available N, and fire-promoting nature of pine, make it possible for the latter species to resist reinvasion by oaks. Since the seed dormancy in a majority of tree species is lacking, most of them may become indeed, nonrenewable.

The total growing stock of forest at 66 x $10^6 m^3$, out-turn at 4.54 x $10^6 m^3$ per yr and annual increment at 0.78 x

$10^6 m^3$. Thus, with a consequent net depletion of 5.8% per yr the growing stock by the year 2041 will be only 3.32 x $10^6 m^3$. Evidently, further destruction of forest must be prevented, pressure on them for fuel and fodder must be reduced and the blanks must be refilled by concentrated forestry programmes.

The permanent well-stocked forest cover is also needed in view of the following: (i) The catchments are subsurface flow systems, hence they are vulnerable to landslides. The tree roots provide (a) deep channels which facilitate the subsurface flow, and (b) a structural biotic net work to hold the soil body in place. Removal of trees disturbs the natural hydrological cycle, and thereby enhances the rate of landslides. because of this property, i.e. subsurface flow, the road-building activities also have to be thoroughly evaluated. (ii) of the total nutrient pool of the forest ecosystem, the biomass accounts for a considerable proportion. This provides a mechanism through which the nutrients being released by the geological weathering and decomposition, are conserved against likely losses due to monsoon conditions.

The forest removal will rapidly deplete the nutrient capital of the system, leading to the loss of soil fertility. Although farm forestry is recommended to replace the marginal and steep-slope cultivation, large-scale monocultures should be discouraged. Our findings indicate that in multistratal, complex and species-rich native forests various growth forms and constituent species have evolved in interaction among themselves, each occupying a niche which is different from those of others in a subtle manner. For example, peak photosynthetic activity of one growth form is separated in time from that of the other. This type of niche-differentiation optimizes the overall productive efficiency of the ecosystem, although a given species may be less productive than the exotic species being introduced in monoculture. On the basis of the rate of litter decomposition and concomitant nutrient release, a number of groups of species are identifiable within a forest. This ensures that the nutrients are not released from the litter in a single annual pulse, and hence they are conserved against leakages from the ecosystem. Mixed species tree culture, therefore, is preferable.

The multistratal forest canopy also protects the soil from the impact of rain drops and thus minimizes the erosion hazard. Comparatively more of the potential, active and old landslides are associated with pine forest of poor canopy, than with multistratal broadleaf forests. While selecting the exotic species for large-scale introduction, the structural and functional attributes of the native vegetation need to be kept in view. The structural and functional features which are widespread in the vegetation today are those which have been selected during the long course of evolution under the prevailing climatic conditions. The exotic species which resemble the native species in regard to functional features are likely to be more successful in natural competition than those possessing different characters.

Kumaun Himalaya comprises the districts of Naini Tal, Almora and Pithoragarh lying between 28°43'55" and 30°20-12" N lat. and 78°44'30" and 80°18'45" E long. On the basis of altitudinal variation the area divisible into subtropical, temperate and alpine zones representing a variety of forest types. These forests constitute the main source of timber, fuel and minor forest products. Apart from the pecuniary value, the forests play an extremely important role in the water and soil conservation. However, due to a sub-optimal forest cover and density, occasioned by a long history of deforestation and mis-management, the protective role of forests has declined. Soil erosion and water runoff in the region have reached alarming proportions as evidenced by silt-laden torrents, streams and rivers during the rains and the denuded and eroded hills which were once carpeted with a thick mantle of vegetation. Unstable geological formations, steep slopes, ever increasing population pressure, deforestation, in-adequate soil protection due to decreased vegetation and litter, overgrazing and lopping recurring fires, destructive mining practices and defective road construction techniques are some of the causes which have led to the present alarming situation.

The present paper is an attempt to collate the available information on the forest types and the associated problems of the region. The Kumaun Himalayan region consists of four forest divisions viz., East Almora, West Almora, Naini Tal, and Pithoragarh. Champion & Seth have classified Indian forests into sixteen groups. Out of these, the forests of Kumaun Himalaya represent seven groups, viz., Tropical moist deciduous forest, Dry tropical forests, sub-tropical pine forests, Himalayan moist temperate forests, Himalayan Dry temperate forests, Sub-alpine forests, and Moist alpine scrub. The dry tropical forest group exclusively belongs to the Pithoragarh division and Himalayan dry temperate forests, the subalpine forests and the moist alpine scrubs are confined to East Almora and Pithoragarh divisions. The moist mixed deciduous forest, the Himalayan chir pine forest, the Ban oak forest, the Moru oak forest, the moist deodar forest, the oak scrub, the Kharsu oak forest and the Alder forests representing the groups, tropical moist deciduous forest, subtropical pine forest and Himalayan moist temperate forests occur in all the four divisions.

The total forested area in Kumaun Himalaya is about 3,38,300 hectares and the division wise break up is as follows:

East Almora Division 1,13,045.8

West Almora	79,086.4
Naini Tal	59,317.3
Pithoragarh	86,850.5

Shorea robusta, a characteristic species of the north Indian moist deciduous forests of the Siwaliks, with Anogeissus latifolia, Terminalia tomentosa, and Dendrocalamus strictus as main associates, occurs in the altitudinal range of 305 m to 1,067 m in Naini Tal division and from 900 to 1,200m in West Almora and Pithoragarh divisions. The area covered by this type of forest in these divisions in about 9,403.3 hectares.

Occurring only in East Almora division with a floristic composition similar to the moist Siwalik Sal forest, on purana formation with light alluvial soil and boulders, covers the elevation range between 500 and 1,100 m, forming a narrow strip in and around the main river valleys.

The moist mixed deciduous forest is mainly composed of Albizzia procera, Adina cordifolia, Terminalia tomentosa, Cedrela toona, Salmalia malabarica, Lagerstroemia parviflora and Grewia oppositifolia, extends from 600 m to 1,500 m along exposed slopes with poor soil, derived from trap rocks, shales and slates. This type occurs in all the four divisions of Kumaun and covers an area of about 1,4354.8 hectares. Confined to the Naini Tal division and occurring as an underwood in the moist mixed deciduous forests Dendrocalamus strictus forms relatively low but often dense brakes on dry hill sides and sometimes on alluvium.

Occurring on the dry and exposed ridges in the Pithoragarh division and covering about 260 hectares, the northern dry mixed deciduous forests are chiefly composed of *Anogeissus latifolia, Bauhinia spp., Terminalia tomentosa, Woodfordia fruticosa, Bauhinia vahilii* etc. extends up to 1250 m altitude or even higher on the southern slopes merging with Pinus roxburghii forests.

Two types of chir pine forests viz., the Siwalik chir pine forests and the Himalayan chir pine forests occur in the Kumaun region. The Siwalik chir pine forest covers an area of approximately 881 hectares in Pithoragarh division and is represented by Pinus roxburghii mixed with Shorea robusta, Teminalia tomentosa, Anogeissues latifolia and Woodfordia fruticosa. In relatively moist places Quercus incana, Q. glauca, Myrica sapida and Rhododendron arboreum also occur together with Pinus roxburghii. The Himalayan chir pine forest is much more extensive covering an area of about 1,93,893 hectares and occurring between the altitudes 750 and 2,000 m in all the four divisions of Kumaun region.

The growth and regeneration of P. roxburghii is better on cooler aspects with deep soil and good drainage. Beyond the upper altitudinal limit, P. roxburghii gives way slowly to the oak forests while towards the lower altitudinal limits the type gives place to either the sal forests or the moist mixed deciduous forests. The recurring fire in the chir pine zone is the most favourable factor for the occurrence of this forest.

Most frequently Proxburghii forms a pure crop in this area but sometimes it is also found mixed with certain broad-leaved species such *as* Q. incana, Q. glauca, Pyrus pashia and Rhododendron arboreum. In the East Almora and Pithoragarh divisions a sub-tropical scrub vegetation composed of species of Carissa, Dodonea, Rhus and Woodfordia, between the altitudes of 600-1, 500 m covers considerable area in the Siwalik chir pine zone extending sometimes into the Himalayan chir pine area.

Q. incana is the most common oak occupying an area of 5,35,44 hectares and occurring in all the four divisions. The altitudinal range of Q. incana is from 1100-188 m on northern aspect and from 1800 to 2100 m on southern aspect. The chief associates of this oak are Rhododendron arboreum, Lyonia ovalifolia, Pyrus pashia, Myrica sapida and Cornus oblonga. In the Naini Tal division between the altitudes of 1,800-2,100 m another oak, viz., Quercus lanuginosa forms local consociations with the help of Q. incana, Rhododendron arboreum, Pyrus pashia, Myrica sapida and Lyonia ovalifolia especially on the southern aspects.

The Q. dilatata forests occupy an area of about 9,925 hectares and occur in all the four divisions between the altitudes of 1,900-2,700 m. Compared to Q. incana this is a more thermo-phobic species growing best on northern aspects and forming a transitional zone between Q. incana forests at lower altitudes and Q. semecarpifolia forests at the higher altitudes. The common associates of Q. dilatata are again Lyonia ovalifolia, Rhododendron arboreum, Pyrus pashia, Viburnum cotinifolium, Cornus oblonga, Symplocos chinensis and Lonicera quinquelocularis.

Especially near the human habitations, due to an accentuated intensity of lopping, Quercus incana and Q. dilatata have assumed the appearance of a scrub with mixtures of Rhododendron, Berberis, Crataegus, Spiraea and Wikstoremia. Confined to the altitudinal range of 1,800-2,150 m in the West Almora and Naini Tal divisions a temperate secondary scrub vegetation representing secondary seres has developed irregularly but rather extensively. The important species are, Plectranthus rugosus, Berberis asiatica, B. lycium Spiraea canescens, Indigofera gerardiana, Crataegus

The only representative of Himalayan dry temperate forests, Hippophae salicifolia forms more or less pure stands, 2-6 m high, along the inner Himalayan streams in the East Almora and Pithoragarh division between the altitudes of 2,300-3,200m. Salix elegans, *Myricaria elegans,* and *Populus ciliata* are sometimes mixed with *H.* salicifolia at lower altitudes while Juniperus is mixed with it at higher altitudes.

The cedrus deodara forest occurring mainly between 1,700 m and 2,500 m altitudes but extending appreciably lower on cool aspects and much higher, even to 3,300 m, on sunny ridges, preferring particularly a drier climate, covers total of about 742 hectares in all the four divisions. The typical associates of Cedrus deodara are Pinus wallichiana, Q. incana, Q. dilatata, Indigoferu gerardiana, Lonicera angustifolia and Rosa moschata.

Moist temperate deciduous forest is commonly found between 1,800 m and 2,750 m altitudes in moist hollows and depressions or as strips along hill streams covering an area of approximately 2,007 hectares in East and West Almora and Pithoragarh divisions. The common species occurring in this area are: Aesculus indica, Acer caesium, Carpinus viminea, Ulmus wallichiana, Betula alnoides, Juglans regia, Fraxinus micrantha, Quercus semecarpifolia, and Taxus baccata, etc.

Occurring between the altitudes of 1,500-3,000 m in the Pithoragarh and Naini Tal and mixed with Q. dilatata, Q. incana, Q. semecarpifolia and Picea smithiana, Pinus wallichiana form local seral communities.

Covering an area of about 1,94,17 and occurring in all the four forest divisions, Q. semecarpifolia is the most important oak of the attitudinal range 2,500-3,300 m with Betula alnoides, Quercus dilatata and Abies pindrow as its chief associates. Several ecologists believe that one or the other of the oaks form climax for the temperate vegetation of Himalaya; Q. incana and Q. dilatata at lower altitudes and Q. semecarpifolia at the upper altitudes. Between the altitudes of 2,600-3,400 m, Q. semecarpifolia forms a mixed forest with Abies pindrow in the east and west Almora as well as in the Pithoragarh divisions. These mixed oak-fir forests cover an area of about 2,701 hectares.

Occurring in the East Almora and Pithoragarh divisions and covering an area of about 4,910 hectares, the Tsuga dumosa forest forms the western limit of East Himalayan mixed coniferous forests between the altitudes of 2,300 m to 3,000 m. At higher elevations T.dumosa is gradually replaced by *Abies pindrow* while at lower elevations a gradual transition occurs to moist temperate forests dominated by oaks, *Acer* and *Betula.*

Cupressus torulosa in the East Almora and Naini Tal forest divisions occurring mostly on limestone rocks covers an area of approximately 1, 687 hectares. The Alder forests are more widely distributed occurring in all the four divisions between the altitudinal limits of 1,000 m to 3,000 m but are predominant only locally. Populus ciliata and Ulmus wallichiana are the most important associates of Alnus. Forming a dense undergrowth in the oak and coniferous forests and often predominant along moist banks of streams, Arundinaria falcata represents the montane bamboo brake in all the forest divisions of Kumaun with the exception of West almora division.

Betula utilis and spectabilis from a mixed forest, sometimes with the help of Quercus semecarpifolia and Rhododendron campanulatum within the altitudinal range of 3,000-3,500 m in the East Almora and Pithoragarh divisions. These forests cover an area of about 3,342 hectares. At relatively higher altitudes i.e. beyond 3,100 m in the East Almora and Pithoragarh divisions there occurs a low evergreen scrub association of Betula utilis, Rhododendron campanulatum. Viburnum nervosum, Rhododendron lepidotum, Berberis kumaonensis, and Juniperus recurva. The vegetation hardly grows a more than 1 metre high, and covers about 325 hectares in East Almora division. Apart from the above forest types extensive pastures are also found in the region. The temperate pastures occurring between 1,800-3,000 m and dominated by species of Festuca, Agrostis and Dactylis are found scattered within the forested areas of East Almora and Pithoragarh divisions.

Between the altitudes of 3,000—3,500m the subalpine pastures dominated by species of Agropyron, Brachypodium, Bromus, Poa and Festuca cover an area of about 896 hectares in East Almora and Pithoragarh divisions. These pastures gradually merge into alpine meadows at high altitudes which are characterised by mesophytic perennial forbs with little grass component. Primula, Anemone, Fritillaria, Iris, Gentiana etc. give a characteristic aspect to these meadows which cover an area of about 2,479 hectares in East Almora and Pithoragarh divisions.

On the basis of a study of satellite imagery of these areas and the existing topo and forest maps, a reconnaissance land use classification of the area, was developed. The larger proportion of total area under forest, though still below the requisite level for hills, as compared to that under agriculture, underlines the basic importance of forests in the general economy and development of the region. The area occupied by important species in the forest under the control of forest department is given in the table 2.

Out of the total forest area in Kumaun circle under the control of Forest Department 55.9% area is occupied by *Pinus roxburghii.* Of the rest 24% is under *Quercus* spp. and 3.4% under *Shorea.* In the civil, panchayat and private forests which comprise nearly one third of the total forest area no forest management has been practised. These forests have been extensively exploited for meeting the ever increasing demands of the local people, without any effort at afforestation and regeneration of the forests. This over utilization of forests combined with uncontrolled grazing has resulted in the disappearance and degeneration of valuable forest wealth making the hill slopes barren. We do not want to enter into controversy with regard to the state of health of the forest sunder the management of forest department. However, from feelings expressed from time to time and from frequent visits in the Himalayan region it appears that all is not well. Our limited study in the Naini Tal hills on certain forests which are under the control of the Forest Department, indicates a range in the total forest basal cover of 1764 cm^2 per $100m^2$ to 7511.5 cm^2 per $100m^2$ in *Q. dilatata* forest and from $2170.8m^2$ per $100m^2$ to 7139.4 cm^2 per 100 m^2 in Q. incana forest.

During the year 1975-76 the maximum amount of timber was harvested and marketed form P. roxburghii : Naini Tal 11547.9 m^3, East Almora $44921.6m^3$ West Almora $35809.5m^3$ and Pithoragarh 21247.0 m^3. In 1976-77 the exploitation of P. roxburghii forests in all the four divisions decreased: Naini Tal 5667.3 m^3, East Almora 3748.0 m^3, West Almora 35100 m^3 and Pithoragarh 20901.0 m^3. Rest of the species with the exception of Shorea in Pithoragarh division and Kukat in Naini Tal and West Almora divisions, were not important as far as marketing of timber is concerned. No data are available on the removal of total plant biomass.

Regeneration is one of the main problems in this region particularly for broad-leaved forests. As soon as protection is afforded from burning and grazing, the regeneration of P. roxburghii become fairly abundant. In the case of *Shorea robusta*, in forests with a broken or open canopy, subject to periodic or annual fires, plenty of regeneration comes up and persists for long periods in the wippy form. Subsequent protection is required to produce right condition for these seedlings to grow up.

In Cedrus deodara, although appearance of seedlings seems to be satisfactory, closure to grazing and full protection from fire are essential for the successful natural regeneration. In high level *Abies* no natural regeneration appears under mature forests which have a dense canopy with a considerable accumulation of raw humus or highly humified upper soil. Opening up of canopy and introduction of light grazing disturb the raw humus, expose patches of mineral soil and also reduce weed growth, resulting into a gradual increase in regeneration.

Greater openings are subsequently required for the further development of the small saplings. In *Picea* regeneration seems to be rather poor. Saxena *et al.* have recently studied the regeneration potential of certain oak forests in Naini Tal hills. In the mixed oak conifer forest the maximum number of saplings were recorded for Q. *dilatata* followed by *Ilex dipyrena and I. odorata*. There were 5.5 saplings 100m^2 of Q. *incana* but no seedlings. *Cedrus deodara* with only 5 saplings/100m^2 showed maximum number of seedlings followed by *I. dipyrena* and Q. dilatata.

In Q. dilatata forests the number of saplings range from 3 to 36 per 100 m^3. In this forest the saplings of Q. incana were found in some stands with a range from 4 to 12 saplings/100m^2 seedlings of this species were recorded only in one stand, with a density of 1.3 per 100m^2. In Q. incana forest the density of saplings ranges form 0.3 to 6.7 per 100 m^2. No seedling of this species was found in any of the stands. In these forests low number of saplings of *Rhododendron arboreum, Cornus oblonga and Lyonia ovalifolia* was recorded and seedlings were either absent or negligible. From the foregoing account it seems that the regeneration potential of *Q. dilatata* is adequate while in Q. incana the possibilities of regeneration are remarkably low. Especially near villages the regeneration of Q. incana is totally absent. For most of the species closure to grazing and burning is essential for successful regeneration.

Forests are the main source of timber and fuel and often of fodder but at the same time they have an extremely important function in water and soil conservation. A part from low percentage of forest area, present density of forest is suboptimal for hills. One of the major causes of the steady depletion of forest cover form this region is the large scale and continuous lopping of oaks and other broad-leaved species for cattle fodder and felling for fuel. Area around Naini Tal contain mostly oak forests which have been badly lopped by the milkmen and other graziers who keep large number of cattle in the vicinity of these forests. The oak forests are malformed occurring as pole groups which look underdeveloped due to maltreatment.

In Pithoragarh forest division under sal forests, the area of Amri, Puraban and Palbelon west block are much deteriorated due to past maltreatment. In East Almora Forest Division at a number of places like Boel compartment number 2b; Papoli 5b, 6b; and Rangorikot 6 etc. Only small patches of miscellaneous forests are left.

A majority of population in this region has agricultural and pastoral occupation. A large number of cattle are kept, more for production of manure for the poor soil than for animal products. However, villagers have rights to graze their cattle in the adjoining forests. Most of the villages have also rights of Kharak, i.e., making temporary cattles stations in forest areas in certain seasons of the year annually. In addition to the local graziers, migratory graziers like Gujjars who are inhabitants of high level valleys, graze their cattle during summer and rainy seasons in alpine pastures, and move to tarai and bhabar tract in winter months.

The grazing affects forests in two ways : (i) physical injury to the plants in seedling and saplings stages decreasing the regeneration and productivity and (ii) planned burning by villagers to produce more succulent herbage. It is estimated that about 5 heads of cattle are maintained to produce sufficient manure for 0.4 hectare of cultivation and to feed them 5 to 6 hectares of forests are lopped and destroyed. Grazing status for the Almora, Naini Tal and Pithoragarh. In Almora district only 4,88,800 hectares are actually available for grazing. Taking the carrying capacity of the forests of Almora district works out to be about 9,77,600 cattle units. But the total number of cattle units recorded in 1961, 1966 and 1972-74 were 28,86,500, 29,85,000 and 22,89,000, respectively. Thus, the incidence of grazing was about 2.95 times higher in 1961, 3.05 times higher in 1966 and 2.34 times higher in 1972-74 than the estimated carrying capacity of the forest.

In the Naini Tal district 2,95,000 hectares are available for grazing. The total carrying capacity of the area works out at 5,90,000 cattle units. The total number of cattle units is respectively. Thus the actual grazing incidence was 3.26, 3.58 and 3.68 times more than the carrying capacity in 1961, 1966 and 1972-74, respectively. In the Pithoragarh district 2,37,300 hectares are available for grazing and as such the total carrying capacity of this area will be 4,74,6000 cattle units. But the total number of cattle units recorded in 1961, 1966 and 1972-74 were 13,95,000, 14,21,000 and 19,60,000, respectively. The grazing was about 2.94 times higher in 1961, 2.99 times higher in 1966 and 4.13 times higher in 1972-74 than the carrying capacity.

Soil erosion and afforestation: In this region the problem of soil erosion is acute. Some compartments of Kuria like 4, 5 and 6, Sat Tal compartment numbers 6b, 10b etc., Khurpatal compartment numbers 1, 2, 3b, 5b, etc., Chanda Devi 7, 8, 9 etc. and Ruinsi compartment number 1 in Naini Tal Forest Division show acute erosion.

Many of the compartments along the main high way to Naini Tal and Ranikhet are being badly eroded. Plantations raised in these areas in the past have failed mostly. Almost whole of the area of Kuria compartment number 5 and part of compartment number 6 which lie above Kathgodam-Ranikhet motor road, has gentler slope with lesser erosion. Similar is the case with Ruinsi compartment number 1 along the Naini Tal road. In the eroded areas which are near habitation, immediate closure against grazing is called for. Subsequently, soil working should be taken up with a view to conserve soil and water. On the steeper areas, mainly above motor road where the soil depth is negligible and is calcareous and rocky, contour furrows of a trapezoidal cross section are recommended. Strips of grasses such as Chrysopogon fulvus, Panicum maximum and Chloris incomplete etc. Can be tried. At places where fine gravel preponderates, staggered terrace may be made with planting of grasses, Agave bulbs, opuntias etc.

On northern aspects containing deep soil having fairly open crop Acacia Dealbata and A. mollissima are recommended. Cupressus torulosa should be planted on shallow soil and rocky southern aspect. On southern and western aspects and on steeper and drier slope with light and sandy soils, Robinia pseudoacacia to be tried. This species has a wide tolerance. On northern and eastern aspects and on easier slope with good deep soil and humus conditions, Juglans regia, Picea smithiana and Abies pindrow and Cedrus deodara are recommended for planting on cooler aspects with moist, well drained and deep soil. The exotic pine trees should be planted on the north and north west aspect containing deep soil. Pinus densiflora is fast growing in comparison to P. roxburghii and is grown at high altitudes.

Gupta has suggested some species for different habitats and altitudes. On contour and field bunds Pennisetum purpureum, on terrace faces Apluda mutica, Arundinella nepalensis, Themeda anathera, Phalaris tuberosa etc. can be planted. He also recommended the plantation of Cynodon dactylon, Chrysopogon fulvus, Bothriochola pertusa, Atylosia scarbaeoides, Morous alba, Bauhinia variegata, B. Purpurea, Cedrela toona, etc. at an altitude of 500-1000 m and Festuca pratensis, Imperata cylindrica, Arundinella nepalensis, Arundinaria spp., Trifolium pratensis, Robinia pseudoacacia, Celtis caucasica, Juglans regia etc. at an altitude of 1,500 to 3,000m on steep slopes, wastelands gullies.

The other species which can be used for plantation include Aceroblongum, Morus alba, Salmalia malabarica, Pinus Wallichiana, Aesculus indica, Alnus nepalensis, Carpinus viminea, Ailanthus sp., Cryptomeria japonica etc. All plantations should remain strictly closed to grazing and lopping by erections of cattle proof fence and should also be fire protected. Looking to the needs of water conservation and rapid build up of soil, as far as possible, broad-leaved forests with multilayered canopy should be raised.

The society of American Foresters defined forestry as the scientific management of forests for the continuous production of goods and services. The forest itself has been defined variously as : (a) plant association predominantly of trees and other vegetation (b) forest lands bearing a tree or bamboo cover, whether productive or not (Food and Agricultural Organization of the United Nations) and the forest land was defined as all land bearing vegetative associations dominated by trees of any size, exploited or not, capable of producing wood and other forest products, or exerting an influence affecting each other and associated plants and animals, and interacting with soil and local climate.

Morosow's definition is more holistic and conforms to the present nation of ecosystem. As the definition of forest varies so do the aims of forestry. While on the one hand, forestry may attempt to maximise the production of merchantable wood or to increase area under a given species on pecuniary considerations through orchard forestry, on the other, it may aim to conserve the total forest environment employing such techniques as naturalistic silviculture which is based on synecological knowledge.

In naturalistic silviculture, according to Bakuzis, species composition and structure can be altered but it is nevertheless adapted to its site. There term 'conserving' here is not meant to be synonymous with 'preserving' but rather means a kind of management which ensures sustained yield but does not deteriorate the environment; it adheres as closely as possible to natural conditions under sound economic planning. It may, however be imperative to 'preserve' a considerably large part of the hill forest for variable time period so as to halt the fast degradation of the ecosystem and to obstruct the changes form becoming irreversible.

The objectives of the forestry should be in tune with the regional environmental needs and considerations. Thus while an European forester may have to deal with the problem of water surplus, an American forester usually faces the problem of water conservation. In the Himalaya today the most burning problem is that of soil and water conservation and evidently it is this function of the forest that has to be maximised through forestry even at the cost of pecuniary losses. The emphasis should perhaps shift at least for time being, from forest crops management to 'watershed' management in Himalaya.

The distinction between forest and watershed management is evident because while ecosystems, at the standard level of forestry, are delineated at a level of forest communities which respond homogeneously to treatment, for watershed or range management delineation at the landscape level is needed; the landscape being defined as a part of the earth's surface with a certain physiognomy, or a characteristic appearance with causal relations and structure. For the watershed or catchment management a land capability classification is important and the unit must be managed as a 'system' on the basis of ecosystem principles. Clearly the maximization of soil and water conserving function cannot be achieved unless the firewood and fodder crises are solved. This will involve large scale afforestation of quick-growing fodder and fuel trees, development of forest grasslands, and provision of energy subsidy in the form of coal and cheap hydro-electric power.

The competition between forestry and agriculture has become acute and this has resulted into cropping on slopes much too steep for agriculture. Gupta estimates about 3,000 Km^2 croppable land in UP Himalaya against 9,000 Km^2 being actually cropped. If Himalayan ecosystem is to be saved an alternative to agriculture has to be discovered. Here orchard silviculture and development of forest based industries might be of immense help.

The forest resources should be linked with forest-based industries and forestry should be developed to support rural economy. A multiple purpose forestry needs be advocated. All of the foregoing would suggest a deep impregnation of forest ecology in forestry. The forest ecological studies must provide a sound base for both orchard and naturalistic silviculture, and for the conservation of total environment from ecosystem viewpoint as against sectoral management of just forest stands and tree plantations. From time to time the management policy adapted for a particular region must be reviewed by scientists other than those involved in policy making and operation, and the ecological effects of management must be determined by comparisons with 'controls' set aside as biosphere reserves. A search in the literature and available records indicates a conspicuous lacunae in quantitative data base for forests.

The statistics on forest land is available rather precisely, the true extent of the forest cover is not known for any of the hill regions. Similarly, although, the volume of marketable timber is fairly well estimated, the total biomass and its partitioning into different components is not known for any area. Rates of forest growth in different types of physiographic and climatic situations are also wanting. Effect of different past management practices on the long term forest growth, stability and regeneration have not been quantified. The relationship between the forest cover and such abiotic parameters as water yield, infiltration, rainfall apportionment, etc. have not been studied. The success of afforestation depends to a large extent on the autecological parameters of involved species. Detailed studies on most of the components of forests are still lacking. Similarly studies on primary productivity and nutrient cycling from the viewpoint of ecosystem approach are prerequisite for developing any rational forest management plan, and so far no such study is on record.

A need has been felt since long for the education of those involved with the forest management and those who ultimately use the forest resources or otherwise are involved in forestry operation of various kinds. The natural question which comes up is as to what should be the content of such an education? Although, we make no judgements, we strongly feel that whatever education is imparted today for such purposes is based on the philosophy so forest exploitation with little input into the programmes of ideas and results originating from research.

Traditionally no university was involved in such a teaching programme. It is encouraging to note that the hill universities namely—the Kumaun and Garhwal and some of the agricultural universities have started offering courses in forestry. These courses must have a very large component of ecology. In addition to the training to the professional forests, short term formal or non-formal courses must be instituted in the universities particularly for the users of the forest resources and for the audience such as forest contractors and personnel of forest corporations in order to inculcate in them the respect for total forest environment. From time to time extended workshops involving experts and workers of voluntary organizations should be organized to sensitize the decision and policy makers.

Landsat-1 images were visually interpreted for the assessment of forest cover in the Indian Central Himalaya. Forests were classified into three status classes, viz, 'good', 'medium' and 'poor'. For each class, biomass ranges were standardized from aerial photointerpreted biomass maps, and biomass for entire area was computed. Kumaun and Garhwal regions had 34.1 per cent about 24.9 per cent of their respective land under forest. 'Good' forests accounted for 4.8 per cent land in Kumaun, 4.1 per cent in Garhwal, and 4.4 per cent in whole of the Indian central Himalaya. In the entire area total above-ground biomass was 30765 X 10^4:

The maximum biomass was in the Naini Tal district and the minimum in the Dehradun district. The biomass density averaged 65.2 t ha^{-1}. The biomass estimations are required especially for the determination of site productivity, nutrient cycling and energy potential. Traditional field methods, involving plot harvests, require massive inputs of time and money and become impractical for large areas. The use of satellite images combined with the aerial photographs could be a viable alternative for the purpose.

Application of Landsat data for resources surveys started with the launching of the first Earth Resources Technology Satellite (later renamed Landsat 1) by NASA on July 23, 1972. Sahai and Samuel used the Landsat images for land use classification of Kerala State, Rao and Unni used them for forest type classification and mapping of shifting cultivation areas in Mizoram, and Unni et al. studied the spectral responses of various land cover types through Multispectral Digital Analysis System (M-DAS). The present study deals with the assessment of forest status in the Indian Central Himalaya using Landsat 1 images.

The Himalayan ranges extending in altitude from about 250 m at the base to >8000m, support a variety of forests below the timber line. The central sector of the Himalaya comprising Nepal, and Kumaun and Garhwal regions of the Uttar Pradesh in India, is called Central Himalaya. Indian Central Himalaya extends from the natural divide of the Kali river in the east, bordering Nepal, to Tons-Pabar valley in the west.

The area comprises Naini Tal, Almora and Pithoragarh districts of Kumaun, and Pauri, Chamoli, Uttarkashi, Tehri and Dehradun districts of Garhwal. The altitudes varies from about 250m to >3500m. According to Manni, by combining the differences in temperature and humidity, a single mountain range can provide a bewildering variety of climates. The rainfall pattern is largely governed by the monsoon, however, the influence of monsoon declines as one ascends higher altitudes. Within the Lesser Himalayan region of Kumaun Himalaya, the annual rainfall tends to increase from 1441mm to 2416mm with increase in elevation from 300 to about 2200m. with further increase in elevation rainfall tends to decrease, and the ranges of trans Himalaya receive conspicuously lower rainfall. There the snowfall becomes substantial.

The mean annual temperature (MAT) in the foot hills (Tarai areas) is about 27°. It declines with rise in elevation and can reach-4 or -5°C in the show covered areas. On the basis of rainfall and temperature, the year is divisible into three seasons, viz, rainy (mid-June to September), winter (November to February) and summer (April to mid-June). October constitutes the transitional period between rainy and winter, and March between winter and summer. The region supports subtropical broadleaf forests below 1200m, pine forests between 1000 to 2200m, moist temperate evergreen broadleaf and mixed coniferous forests between 1500-3500, and sub alpine forests around 3500m.

The study was performed using Landsat-1 images for the year 1972. False colour composite of bands 4, 5 and 7 were used. The area was covered by three Landsat scenes in two orbits (path No. 157: row no.39, path no.159: row no. 39, and 49). The false colour paper prints on 1:250,000 scale were procured from the national Remote Sensing Agency. The black-and-white aerial photographs for the year 1973, which were available for a smaller area, and survey maps served as the 'ground truth'. Visual interpretation technique was used. Briefly, three status classes of forests were distinguished. The area with absolutely uniform red colour was considered 'good' forest. Examination of the corresponding aerial photographs for some of these 'good' forest areas indicated that these forests have a general crown cover of >60%. The comparatively lighter red colour with a few scattered yellow dot-like structures was considered 'medium' forest, and these loosely corresponded with forests having 30-60% crown cover. The area where red colour was dull and was conspicuously mixed with yellow was considered to represent poor' forests, which had 10-30% crown cover.

Non-forested land was classified into snow(appearing white on the images), high altitude meadows (appearing brownish yellow near the snow line) and other non-forest uses which included agriculture, waste land, grasslands, eroding lands, barren lands or denuded hills (appearing yellow on the images). A base map of whole area having district boundaries was prepared on a transparent sheet on 1:250,000 scale. The base map was placed over the false colour composite print on a light table. Boundaries of different land uses and forest qualities were marked on the base map using a magnifying lens. The area of different land uses and forest classes was estimated using a dot-grid overlay.

A method for mapping of forest biomass using aerial photographs and non-destructive field sampling was developed. Using this method, the total aboveground tree biomass maps for 10 subcatchments of Kumaun Himalaya were prepared. The above-ground biomass ranges for 'poor','medium' and 'good' forests were standardized using these maps. Interpreted biomass ranges were 80-160t, 160-280t and 280-480t per ha, respectively, for 'poor', medium' and 'good' forests. By multiplying the area of different forest status classes with mean values of these ranges, total aboveground tree biomass for different districts was computed.

The forest vegetation in India which according to Hooker is the richest and most varied on the surface of the earth is governed by a complex of environmental features, including climatic and edaphic factors and varied topography. Owing to wet ecological conditions ranging from tropical forests to temperate belt in the eastern Himalaya the richness of the forest flora in the region is well marked. Excellent accounts of the vegetation of the eastern Himalaya inclusive of subsidiary hills around Khasia and Jaintia region and upper Assam exist in literature. Reference may be made to the notable contributions of Cowan and Cowan, Cowan, Champion, Bor. Rowntree, Chatterjee, Schweinfurth, Puri, Rajkhowa and Champion and Seth. However, in the present context the detailed floristic account of the various forest types around Darjeeling, Shillong plateau (Khasia and Jaintia hills) and upper Assam hills were made with an aim to augment more information on the sites not surveyed earlier and also to document more data on the number and distribution of species among the forests previously studied.

	Place	*Latitude Average*	*Longitude altitude (m)*	
1.	Darjeeling (Eastern Himalayas)	27° 30′N	88° 18′E	2,044
2.	Shillong (Khasia Hills)	25° 40′N	91° 55′E	1,500
3.	Digboi (Upper Assam)	27° 30N	97° 80E	500

The excursion to these localities were undertaken yearly for five years. The localities visited fall within an altitude range of 100-3500m. Summarises the location of the places where temporary research stations were set up to survey the area. The classification of vegetation and also the style of description adopted in the text follows

Champion and Seth. However, the species are listed in the order of frequency in which they are met with.

Champion and Seth divided these forests into seven groups, out of which three viz, tropical wet evergreen, tropical semi-ever-green and tropical moist deciduous are represented in the areas presently surveyed. But opinions differ on its impact on, 'ceeping' slopes. report greater creep velocities in slopes covered by trees in the region of queensland (Austrialia) than in slopes merely covered by grass in a region of rain forests. Brown and Sheu using mathematical analyses based on experimental observations (by other authros) conclude that removal of overload of vegetation and consequent elimination of wind action on the vegetative cover improve stability are temporary and disappear with time. Prandini et al. have presented an excellent state-of-the-art on this subject.

The authors support the views of Croft and Adams, Sares et al. and Poncano et al. that deforestation leads to loss of mechanical support imparted by root system. Reinforcing power of roots is also demonstrated by the results of in situ block shear tests which show that shear strength increases with increase of root density. Besides above vegetation provides benefits of evapo-transpiration, elimination of which would almost certainly tend to raise the ground water table. An individual tree can transpire as much as 50-100 gallons of water on a sunny day and a forest community may stand to lose as much as 50 cm of rainfall in a year depending upon environmental conditions.

What is more, elimination of root system takes away the suction effect in soil which otherwise offers stabilising effect. Plantation of cardamom or cultivation of rice holds water thereby augmenting percolation and hence avoided on unstable slopes. The stability of Himalayain slopes have been to a large extent jeopardised by construction of a network of roads which invariably involves steep and deep back-cutting into hills. These cuttings involve tremendous relief of stress, setting in a process of progressive failure. Due to paucity of funds, it is seldom possible to protect the freshly cut slopes or provide effective surface and subsurface drainage.

The landslides are, therefore, indiscriminate felliing of trees or by tampering with the natural drainage of the area. Human activity is expected to intensify to an unprecedented scale, if 'Dastur Plan' aiming at harnessing of immense water resources of Himalayan ranges is implemented at an estimated cost of Rs. 150000 million to boost India's irrigation potential sixfold and to solve the perennial problems of flood, drought, energy, food and unemployment. The work, if and when implemented, would necessarily involve very wide and deep cuttings which may mean, inter alia, a dreadful onslaught on the ecology of region.

To command landslides is to command nature, which must be obeyed as a rule. The most important of the factors is that natural drainage should never be obstructed or blocked by human activity. On the contrary, all efforts must be made to take advantage of natural drainage courses in improving the overall drainage of the area. Drainage schems for Himalayan Landslides, like most others, involve catch water and contour drains for surface drainage and 'horizontal drains' for subsurface drainage. The Technology for installation of horizontal drains has been perfected by the Central Road Research Institute, New Delhi. These are, however, expensive.

Application of 'bamboo drains' for subsurface drainage may be regarded as a significant Indian contribution that makes effective use of readily available, rather unexpensive 60-80 mm diameter bamboos. These are provided with perforations every 10 cm and driven as far as they go into the slope. Depths of 3-4 m are common. The drains require frequent cleaning and replaeement every third monsoon season.

It is convenient to classify the problms of mass wasting in relation to their gravity. Catastrophes following cloud bursts, flash floods and earthquakes hit headlines in the press, capture public imagination, throw fresh challenges and stretch the exchequer to the extreme. Landslides, shooting boulders and massive rocks transported by the turbulence of flowing water usually trottle the narrow river passage eventually building up a reservoir of water pressure.

The mammoth problems of Himalayan landslides for effective control would require enormous financial resources which would never be available. The wisdom; therefore, lies in fullest exploitation of formidable human resources and in making best use of local skills and materials. A culture of 'preventives' need to be introduced because every landslide prevented or cured means lesser hazard to life and property, less interrupted flow of essential supplies, higher quality of life and greater productivity. Also, the amount of money required in 'preventing' a slide is only a small fraction of that otherwise required in correcting it.

A stitch in time saves nine. Bamboo check dams are found to be useful in arresting further erosion of depressions, channels and gullies on the denuded landslide slopes. Bamboo stakes or stakes of self generating branches or tillers of Ficus cunea, willos, Erythrina species. Phragmites karka, Lannea grandis. Arundo donax et., are useful.

The author treated two denuded slope surfaces of Idgah slide, Simla. 186 sq m and 140 sq m respectively.

using coir netting enough for vegetation to grow without letting the surface soil. seeds and roots to be washed away due to erosvie action of rain.

The length is of retaining walls and breast walls destroyed in the Himalaya during the past decade may add to several thousand metres. These have often been seen to move downward and outward or rupture and sink—being founded on unstable slopes under going large movements. The greatest lesson learnt from the countless failures is that the walls should allow free drainage of water and they should be flexible enugh to absorb ground displacements. Consequently, wire netted 'sausage walls' identical 'gabions' have become popular. Local skills in the art of wire netting are tremendous and the technology offers good scope for generating employment.

Blasting of boulders and slipped masses that throttle the river courses, river training works, prestress-anchoring or rock bolting of slopes and wire netting. Fixed and free fences for rockfall control are some of the other commonly used measures which have already been discussed by several authors and would not be repeated here.

The asphalt mulch method of slope treatment is considered as a promising method of erosion control. When used in combination with other techniques of landslide corection. It helps effectively, preventing surface movements. The brief details of the technique at used at Snowdon slope are given below:

Firstly, the denuded slope area was demarcated and fenced by local prickly bushes and barbed wiring. The slopes were then prepared into vast seed beds, by grading the slope and raking the top soil about 25 mm thick. The slope in question being entirely raw and infertile and the soil being slightly acidic, calcium ammounium nitrate was applied at the rate of 50 kg per 500 sq. m. The root slips of the suitable types of locally available grasses were dibbled. 15 to 22 cm apart. root to root. row to row. Asphalt mulch of a specified grade was then spread by a suitable sprayer. The optimum rate of application of the emulsion was 0.15 to 0.20 gal (1 gallon-4.54 litre) pre sq m.

The thickness of the emulsion coating was regulated as thicker aplication was feared to retard the growth of plants and seeds, whereas thinner application was believed to be ineffective in controlling erosion. Te advantages resulting from the application of asphalt were 9a) the moisture content as well as nutrients in the soil mantle were conserved to promote vegetation. and (b) susceptibility of slope to erosion was reduced.

The asphaltic film eventually disintegrated and its place was graduall taken by a crapet of grass. The method proved particularly successful because it was timed to take advantage of the increased moisture content in the soil derived from first couple of monsoon showers. A comparison showing the untreated slope and showing the same slope after treatment illustrates the benefits of asphalt mulch technique.

An adjoining hill slope, which was also denuded, was demarcated and fenced by prickly bushes. The hill slope was graded to the extent feasible, and all gullies, rills and hollows were filled by soil and tamped firmly. The grading and levelling work was undertaken by manual labour by means of spades, axes, rakers and belchas to facilitate even placement of netting. At the shoulders of the slope trenches were dug out for anchoring the coir netting by embedment, stampling an backfilling. After grading the slope, the soild was raked 12 mm deep by means o hand rakers. Seed mixture was then broadcast over the slide area at the rate of 7 kg per acre and raked into the soil 6 mm deep. To ensure even sowing of the seed, the area was divided into parts and the seeds at the rate of 7 kg per acre and organic manure at the rate of 1000 kg per acre were allocated and two sowings were done, one before and the other after laying of coir netting. The soil was tamped with hand rammers.

The coir netting was firmly laid on the prepared slopes in the direction of water flow. The toe ends of the coir netting (12 mm opening mesh 35 kg per role of one metre width and 50 metre long) were inserted with a 15 cm loop steel staples (u-nails) at 30 cm intervals. The successive widths of coir netting carpets were secured against displacement by an overlapping of about 8 cm. They were pegged down with staple driven 30 cm apart.

The coir netting was fully strectched flush to ground. Thereafter, a does of calcium ammonium nitrate was applied to the finished surface at the rate of 1.5 kg per 100 sq m area together with a second application of seeding at the rate of 7 kg per acre. Besides seeding, the entire area was sprigged with grass spcies. Himalayan landslides are attributed to the combined effect of intensive human activity, excess rainfall, cloudbursts, flash floods, and slope undercutting by meandering rivers, deforestation and seismicity of the region. These have been tackled to varying degrees of effectiveness ranging from spectacular successes to dismal failures. Landslide problems of the Himalaya are complx and defy simple solutions.

The commonly employed corrective measures include measures of erosion control, benching and grading of slopes, subsurface and surface drainage schemes and restraining structures. The effective cure to the problems of Himalayan landslides lies in exploitation of local skills, materials and resources to their fullest potential. Corrective measures involving applications of bamboo, jute and coir netting, wire netted sausage walls are called for. The ultimate solution is bielved to lie in making best use of local resources, skills and of the technology that cuts down both the time and the cost of construction. Finally, Himalayan landslides are not totally avoidable or controllable and a more realistic approach would be to do all that is possible to minimise the gravity of the problem.

Rain forests form an important natural resources of the Indian subcontinent. This resources is shrinking at a rapid rate. Conservations are due to massive exploitation for industrial and fire wood needs, encroachment by the landless peasants, and bush-fallow rotational agriculture (jhum). The present paper considers these conservations and looks into possible remedies. The north-eastern hill region is characterized by a wide variety of climatic conditions. The natural vegetation types range from the humid sub-tropical forests and humid montane forests to alpine meadows. Much of this region receives an annual average rainfall of 2000 mm or more.

The low elevation forests of the hills and that of the Assam Valley form an important segment of the tropical rain forests of the Indian subcontinent alongwith the Western Ghat forests of the peninsular India. The high elevation sub-tropical montane forests are also rain forest types with rich species diversity forming an unique ecosystem complex which needs to be conserved. These luxuriant forest types are often supported on highly leached nutrient-deficient and acidic, soils; these systems are, therefore, fragile. Several factors have contributed to the degradation of these ecosystems. The aim of this paper is to consider these factors and suggest possible remedial measures.

The prime tropical rain forests of north-eastern India are distributed in the Assam valley and long the foot hill regions. In the upper Assam and Cachar hills adjoining to it are found some of the most luxuriant Dipterocarp forests of the sub-continent. In the lower Assam valley, in the Manas region in Assam and extending up to the border of West Bengal and in the adjoining foot hills of Bhutan are luxuriant mature rain forest types of rich species diversity and with none of the important species attaining complete dominance.

At higher elevations are sub-tropical montane rain forests of Quercus spp., Acer laevigatum and Rhododendrons. In Meghalaya they are restricted to sacred groves. Elsewhere they occur in remote areas. While these are the important climax types, there are various stages of secondary successional forests.

Mature rain forest types in north-east India with names of dominant trees

(a)	Upper Assam	Dipterocarpus pilosus, D, macrocarpus, Artocarpus chaplasha, A. integrifolia, shorea assamica and others.
(b)	Cachar Hills	Dipterocarpus turbinatus, Artocarpus chaplasha, Amoora sp., Dysoxylon sp. and others.
(c)	Lower Assam	Pterospermum acerifolium, Mansonia dipikae, Eugenia sp; Dysoxyeun binecteriferum. Phoebe sp; and Amoora wallichii.
(d)	Lower elevation of	Meghalaya Amoora wallichii, Artocarpus chaplasha, Castanopsis indica, and Shorea robusta.
(e)	High elevation montane forests	Quercus sp; Rhododendron arboreum, Acer laevigatum, and Castanopsis sp.

Any natural forest ecosystem is subject to a wide variety of disturbance. These disturbances may be different kinds. Some of these do not necessarily arise from human activities, e.g., death of old trees, wind blow or insect pests and diseases. Even human activities were of very minor dimensions in the past. These minor disturbances may result in gap, building and climax phases forming a mosaic such that the climax phase of the ecosystem forms a major component.

Large-scale exploitation of forests by industrial man results in drastic reduction in the extent of the climax phase or even its elimination resulting in an unstable ecosystem. *Of the reported 21,040 km^2* of evergreen rain forest, 8340 km^2 of semi-evergreen rain forest and 102,000 km^2 of moist deciduous forest for India, a major proportion is confined to the north-east; a narrow belt is along the western Ghats and some in the Andaman and Nicobar islands. Large-scale timber extraction has been done from this region during last few decades. In fact, a large part of the timber and plywood requirements for the country as a whole is being met with from the humid tropical forests of the north-east. According to the 1971 census data for the north-east, the land area under forest was 48%.

Allowing for further exploitation of forests in the past decade and for the fact that the forest cover computed in census reports also includes various stages of degraded communities, perhaps, it would be reasonable to assume that the actual area under good forest cover is only half or two thirds of that estimated in table.

Area under forest in the north-eastern India
(1971 Census Report)

State Area (km^2)	*Land under forest*	*(%)*
Arunachal	83,578	62
Assam	78,523	36
Manipur	22,356	68

Meghalaya	22,489	38
Mizoram	21,087	44
Nagaland	16,527	17
Tripura	10,477	57
North-Eastern State (Total)	2,55,037	48

One of the methods of timber extraction is clear-felling of trees over vast tracts of land. This has adversely affected the forest ecosystem of the mature phase. Another common practice is selective extraction of timber of large size classes of certain species over a restricted area of a few square kms as a "canopy lifting" operation, as is done in some parts of the upper and lower Assam regions.

Though the intention is only to extract larger trees, in practice it almost amounts to clear-felling of forests. The larger trees as they fall down, destroying everything else on their path create much larger gaps than intended. Further, much of this operation is done without giving any consideration for management of forests on the basis of sound scientific studies. Thus, the possible implications of this in the regeneration of trees and the consequent stability of the ecosystem on a long-term basis and the possible impact of this operation on the ecosystem function is not known. Further, since until recently, commercial operators have been interested only in a few selected species, often less than 20 out of 400 species or more present per ha, many of the remaining trees are damaged beyond recovery.

The annual available timber from two of the north-eastern states of Manipur and Tripura is shown in table 3. While these projections give only an indication of the magnitude of timber extraction from these states and the region as a whole, no comprehensive data are available on the magnitude of damage to natural forest ecosystem types.

Annual availability of forest products from Manipur and Tripura (Units in $m^3 \times 10^3$ bamboo in $t \times 10^5$)

Product	*Manipur*	*Tripura*
Saw timber	10.0	7.1
Plywood	110.0	19.8
Pine wood	23.9	—
Hard wood for industry	147.0	—
Pine wood for industry	9.1	—
Pulp wood	—	66.0
Poles	—	1.8
Match wood	—	1.6
Bamboo	1287	5.5

Much of the forested area in the north-east is a degraded bamboo forest of various types. This is related to the shortened jhum cycle (intervening length of the fallow phase between two cropping on the same site) as shown by our studies or even reduced to a weedy community as shown through this study and other studies.

Bamboo alongwith other pulp wood trees is being exported out of the region both for industrial and non-industrial uses. The establishment of three giant paper mills in the region would create a greater demand on the bamboo resources of the region. Unless a sound scientific management policy is evolved now, the bamboo resources would be in short supply in the near future. With the three giant paper mills coming into this region and with bamboo resources dwindling rapidly during the next few years, unless well thought out strategies for regeneration and management of pulp wood trees and bamboo species are implemented soon, pressure on hard wood species from the tropical moist forests would increase. This is particularly so when forest technology has recently developed methods for pulping tropical hard woods in a single mixture.

Much of the firewood consumed in this region,

however, comes from secondary forests. Only a small part comes from primary rain forests. In India, it has been found that from the peripheral zones of the forest boundaries about 70% of the firewood requirements are met with, and beyond 10 km the use of firewood steadily decreases until at about 15 km it is almost nil. However, the increased population pressure and rapid depletion of primary forest cover in the region is rapidly changing this situation. The rural tribal population of the north-east are chiefly dependent upon fuel wood for energy for cooking though other items such as rice husk may also be used.

The fire wood consumption pattern in Sethliew, a small village in the Khasi Hills is given in table 4. In this village of 20 members, 3060 kg of fire wood (52455 MJ of energy) is imported each year, accounting, i.e., 5033 kg fuel wood per year. While rice husk is used as a source of energy here, there are areas where this may not be used and the villagers are exclusively dependent upon fuel wood. On the other hand, there are also areas where fossil fuel in the form of coal is used.

Annual fuel consumption for cooking in the Syth Liew Village ecosystem

Category	*Quantity used (kg)*	*Energy equivalent (MJ)*
Fire wood	5033	86276
Rice husk	328	6032
Total		92308 (92033)
Fire wood imported	3060	52455

The dependence of the rural population on adjacent forests for fire wood is substantial. Consequently, there has been a fast depletion in the fuel wood resources of the region. Methods that increase the efficiency of utilization of fire wood should be employed, for it is low in the developing world. Here the per capita consumption of energy for cooking is considered to be between 2 to 2.5 time more than in the west.

If about 5000 kg of fuel wood were considered to be the average annual consumption by 20 individuals, according to our Sethliew village study, the annual consumption of fuel wood in the villages for a population of 19.58 million (according to 1971 census) would work out to about 489 X 10^7 kg. However, it should be realized that the population has gone up by 30% or more during the past decade and that the fuel wood resources have simultaneously gone down drastically resulting in shortages and high cost for this resources. In order to meet their fuel wood requirements, the villagers are now forced to go out of the village boundary up to a distance of about 10-15 km.

The conversion of rain forests to commercial tea plantation in Upper Assam valley and in the adjoining Cachar Hills was one of the major agricultural operations which started in 1834. In 1961, the area converted to tea was about 178 x 10^3 ha and by 1978 the area was about 250 x 10^3 ha. This accounts for about 53% of the total area under tea plantation in the country.

Forest farming variously termed as shifting agriculture, slash and burn agriculture or rotational bush fallow agriculture (locally called jhum), as practiced in the north-east India is described in detail elsewhere. This land use is a well known practice in the rain forest areas and has survived for thousands of years with little damage to the forest ecosystem. In fact, this farming practice with a long intervening fallow period between two successive cropping (henceforth referred to as jhum cycle) of 20-30 years or more has been in harmony with the environment giving reasonable economic returns of varied kinds (cereals, legumes, vegetables and even fibres) to the farmer. Such a long cycle had maintained a balance between the gap, building and mature phases of the rain forest ecosystem. However, the long jhum cycle was possible partly because there were few farmers per square kilometer, mostly less than 5.

If one looks at the population trends in the north-eastern hill states, one finds that in most of the states, with the exception of Tripura, population has slightly more than doubled in the first 50 years since the turn of the century. In Tripura during the same period it almost went up four times mainly due to influx of people from the plains. Subsequently, in most of the States, the population almost doubled in only 20 years between 1951-1971.

Under these conditions, it is not difficult to visualize the impact such a large agricultural population is likely to have on the rain forests. According to Myers, there are atleast about 140 million forest farmers in the humid tropics of the world. With an average of seven members per family this would work out of 20 million families. If each family clears one additional hectare of forest each year, the total cleared area would work out to 200,000 km^2 each year. According to this estimate, perhaps half of this would be primary forest (about 1% of the total). When considered in conjunction with timber harvesting and other human activities, Myers estimates that almost 200,000 km^2 of primary humid forest is cleared every year which would work out to about 40 ha per minute.

In the north-east India, jhum covered about 425×10^3 families according to 1971 census data. At a growth rate of 24 to 30% per decade, this now stands at about 547×10^3 families and is expected to go to 766×10^3 families by the turn of the century (North-eastern council, 1982). If the projections given by Myers are considered for the north-eastern situation, 0.55 million families at present clear about 6000 km^3 of additional forest for jhum of which 3000 km^2 would be primary forest.

Projection of tribal families ($x10^3$) practicing Jhum

Region	*1974*	*1984*	*2000*
Arunachal Predesh	81	107	155
Assam	58	74	101
Manipur	50	65	92
Meghalaya	68	85	115
Mizoram	45	54	70
Nagaland	80	106	155
Tripura	43	56	78
Total	425	547	766

The exact area under jhum in the north-east is not known at present but recent efforts through satellite imagery should be helpful in this regard. However, the 1974 estimates for this region based on the total area owned by different villagers practicing jhum are given in the above table. According to this, the average jhum cycle works out to about 5 years.

Mizoram has an average of 10 yr cycle. Arunachal pradesh has been considered to have a 3 yr cycle. However, such an average is misleading as the length of the cycle is dependant upon the dispersion of the population. In the same geographical region one could find a long cycle of 30 years or a very short one of 3-5 yr too. Realizing that the data so calculated is an under-estimate of the actual area under jhum, since considerable area is desertified and is no good for cropping and this computation does not consider that rendered useless every year, the magnitude of the problem is enormous indeed.

Much can be said about the scientific basis for shifting agriculture and it is not the purpose of this essay to go into it. However, what propose to discuss here is the feasibility of maintaining jhum in the north-eastern region under a relatively longer cycle of 10 years or more. In a study of the socio-economics of jhum at lower elevations of Meghalaya it was shown that the net gain to the farmer under a 30 yr jhum cycle is more than twice that under terrace cultivation, and that a 10 yr cycle is also economically more viable than terrace cropping under comparable situation. It was further suggested that there is much scope for more intensive valley cultivation, which is a sedentary form of agriculture with steady and sustained yield.

Similar results were also obtained by Mishra and Ramakrishnan for the high elevation Jhum. The cropping pattern under jhum varies under different jhum cycles. With shortening of the jhum cycle, the emphasis by the farmer often shifts from cereals to tuber crops. Under a low fertility system as at high elevation, the shift to tuber crop, such as potato provides better returns to the farmer. The studies on energy budgets on Jhum done at low elevation and at high elevation show that jhum under all cycles, both long and short, is highly energy efficient compared to terrace cultivation.

Apart from the gross energy efficiency the inputs into the system also differ qualitatively. For jhum it is chiefly labour energy while terrace cultivation depends on input of fossil fuel energy in the form of inorganic fertilizers. The generally low energy efficiency of the high elevation Jhum with respect to shorter jhum cycles and terrace cultivation is related to heavier input of organic manure and inorganic fertilizer into these systems which are absent at low elevation jhum systems.

The energy efficiency of jhum cycles cannot be considered in isolation without a discussion of the land use pattern, or else the conclusions would tend to be distorted. If land is not a limiting factor, the greater solar input into a larger area of the jhum system with a longer cycle would offset imported energy and this would ensure harmony of the long cycle with the environment, at the same time ensuring rational returns to the farmer.

With shrinkage in land area available for jhum, the longer jhum cycle can be sustained only if land available for jhum is not limiting. It should also be conceded that under different jhum cycles, the effective output per hectare (based upon the frequency of cropping after different fallow periods) would decrease due to a correction factor of 1/30, 1/15, 1/10 or 1/5 under 30, 15, 10 or 5 yr cycle. Our studies suggest that even on this basis a 10 yr jhum cycle is the most efficient in terms energy ratio and land use. Nutrient losses from the system through run-off and percolation are rather heavy during cropping; the data for N, P and K are given in table for a low elevation system. The magnitude of the losses become obvious only when the correction factor of 1,3 and 6 are applied to the three jhum cycles.

Similar conclusions apply to the high elevation jhum too. Detailed studies on fertility status of the jhum plots under different jhum cycles for low elevation jhum and high elevation jhum are available. These studies suggest the generally low fertility status of jhum plots under short jhum cycles of 5 years as compared to longer cycles of 10 or more years. However, we would like to discuss at this point only the nitrogen economy under different jhum cycles at high elevation which in a large measure would also apply to other nutrients of the jhum systems.

The quantity of nitrogen is much lower under a 5 yr jhum cycle compared to 10 and 15 yr. A lower nutrient level under a short cycle is seen both at the pre-burn stage and at the end of cropping. During the second year of cropping under a 5 yr cycle, the nitrogen capital was even lower at both the pre-burn and after-cropping stages. However, the net loss from the system during one cropping was more under long cycles than under a 5 yr cycle. In the present study on nitrogen budget the plots under a 5 yr cycle had the same cycle length for the past 20 years.

Assuming that these plots had longer cycles prior to this 20 year period, the system has lost, in terms of initial capital, about 1.28 x 10^3 kg of nitrogen (7.68-6, 40 x 10^3kg) during the last 20 years. Thus, while 15 or 10 year jhum cycles are long enough to restore the original nitrogen status in the soil before the next cropping, it is unlikely that about 600 kg of nitrogen lost (the difference between the nitrogen capital before and after two croppings) could be restored under a 5 year cycle. One of the important disadvantages of the 5 year cycle lies in the reduced nitrogen capital with which the agroecosystem has to operate under increased frequency of fire followed by cropping with too short a jhum cycle (every 5 years under a 5 yr cycle as compared to every 10 yr under a 10 yr cycle).

Our studies on nutrient cycling during fallow regrowth at low elevations and at high elevations suggest that there is a rapid transfer of nutrients from the soil pool to the vegetation compartment during the first ten years or so. It is subsequent to this that the recovery of nutrients in the soil pool takes place as a result of the transfer of nutrients through litter fall. Simultaneously substantial additions due to nitrogen fixation and mineralization also contribute to the recovery process.

The efficiency of nutrient cycling in a low elevation jhum system estimated as fractional annual turnover (the weight that left the compartment divided by the weight held in that compartment, expressed as percentage) of the different elements. While the annual percentage turnover of all elements in the soil tended to increase with fallow age, those for vegetation decreased. This is to be expected because of the marked increase in uptake by the developing vegetation and increased storage in the living plants which is higher than the rate of release through litter. The only exceptions to the pattern are for phosphrous and potassium which reached this maximum annual turnover in the soil in the 5-10 yr old fallow.

The high fractional turnover value for phosphrous in the soil in the 5 yr old fallow may be due to a higher uptake by the herbaceous vegetation that predominates there and that for potassium in the 10 yr old fallow may be related to rapid growth of bamboo (Dendrocalamus hamiltonii) at this stage, which stores much of potassium in the living biomass. A major difference between the jhum at higher elevation and that at lower elevation lies in that in the former it has to function under conditions of high soil acidity derived from the Khasi pine needles (Pinus kesiya) which also is a slow decomposer.

4 Flora and Fauna

The distribution of Mediterranean flora has been mainly influenced by two types of climates: (1) Dry desert belt from Sahara across the Arabian Peninsula and Iraq, and (2) The Sub-tropical climatic belt running from Turkey, Oersia, and Afghanistan. The region of Kashmir forms a transition with the Mediterranean and monsoonic regions of Himalayas and this provides the floral limit for the Mediterranean elements. However, the distribution of certain species, east to Kashmir, shows resemblances to Mediterranean types of elements. Some plant species of west Himalayas, particularly the tree-forms show remarkable relationship with those of he Mediterranean climate. Himalayas, a young system of high mountains, situated between Indian Peninsula and Central Asia, connect mountains of the near east and Central Asia with those of East Asia.

From phytogeographical point of view, it is a transition zone between palaeotropic and holarctic kingdoms. In the east, they border the sino-Japanese region whereas at its west end they touch the Irano-Tourranean region which in the widest sense is a part of the Mediterranean territory. Similarity of the climatic regime and phyto-climatic approach has recently been discussed though much a resemblance in the general landscape and plants was observed as early as in 326 BC when the Greeks came to India.

Mediterranean climate has been characterized by a mild winter that does not effect the vegetation, having rainfall during spring autumn, a dry period in summer that checks growth of herbaceous species and stunts the woody ones. However, Emberger laid more emphasis on winter rainfall. Analogous bioclimatic types of the Western Himalayas and chains towards. Alps and Pyrenees have been earlier studied by Gupta. There is no consensus on the eastern limit of the Mediterranean climate is no consensus on the eastern limit of he Mediterranean climate in the Himalayas.

The western boundary of Valley of Indus, having enclaves in Kashmir Valley is considered to coincide broadly with the limit of the Mediterranean climate. However, it is just probable that behind the Himalayan chains, certain valleys may be protected from the influence of monsoon and closely resemble the Mediterranean climate. For such an analysis, a detailed study of each valley would be necessary. Furthermore, the distribution of Mediterranean plants in such a situation would throw more light and would be an indicator for introduction of some economically important species for soil and water conservation in the region.

Vegetation of Western Himalaya and its affinities with that to Mediterranean regions. For analysing the problem of Mediterraneity, the Indo-Pakistan sub-continent has been sub-divided into four physiographic divisions. These are : I. The north-western hill. II. Baluchistan III. The Salt range of Punjab. IV. The Kashmir valley. The natural vegetation of Baluchistan is much disturbed by man. The chief conifer is Juniperus macropoda forming open forests while other trees are Pistacia khinjuk, P. munica, Olea cuspidata, Fraxinus xanthoxyloides, Acacia modesta and Cratageus oxycantha. Pistacia khinjuk and P, *mutica* are the geographic variants. P. vera, P. lentiscus, P. terbinthus and P. altantica of the Mediterranean region and Olea cuspidata that of O. europaea so typical of Mediterranean regions. Nerium indicum extends its range from Mediterranean region to N. W. part of the sub-continent. The stations such as Nokkundi, Dalbandin, Ormara and Pasni in Pakistan support scattered shrubs and thus have been classified as Mediterranean.

Suleman range and Kharewa hills landscape show analogy with that of Mediterranean region. In Kashmir, forests of Quercus ballot and Pinus gerardiana dominate at heights of 1,200 to 2,500m, the former being closely related to Q. ilex of Mediterranean region. Individuals of Olea cuspidata spp., Pistacia, Punica grantatum and Fraxinus xanthoxyloides are *also found. Above 2,250 m* Cedrus deodara succeeds *Pinus gerardiana* and at higher elevations, Abies pindrow A. webbiana and Picea morinda-representatives of temperate forests appear; broad-leaved species like Quercus incana, Acer caesium, Aesculus indica and Ulmus wallichiana thrive in moist depressions. Beyond, Chitral, the vegetation resembles steppes of Central Asia, Waziristan in the north is also recorded to support he Olea cuspidata forests having almond, juniper, and pistachio.

Montane vegetation of shrubs and semi-shrubs has been recorded to form natural communities in Kashmir valley. These grow on volcanic rocks on deeply eroded slopes of Karewa sediments. In some places, it seems to be a secondary formation in place of dry deciduous forests.

A number of workers have shown that in the Kashmir valley there is only a slight influence of the monsoon and the rainfall is concentrated in cold winter season unlike other parts of Western Himalayas. A few species such as Celtis caucasica, Pinus griffithii and Pyrus pashia are present in the higher regions of Kashmir while scrubs such as Controneaster nummularioides, Daphne cashmeriana, Idiogofera gerardiana, Jasminum humile, Plectranthus rugosus, Rosa Webbiana are the low shrubs. Semi-shrubas such as Rumex hastatus, Artemisia, Astragasluls, Polyacanthus, and Lespedeza are semi-scrubs.

Among the perennial herbs, Nepeta sp., Salvia, Saussurea albescens, Trigonella emodi, and several grasses are associated. The scrubs differ in having a large number of extra-tropical semi-arid taxa distribution and evolutionary centre. Some of the species have taxonomic affinities with the West Asian-south Siberian or East Asiana Steppe formation. Flora of the Kashmir valley is relatively poorer as compared to other parts of the west Himalayan flora, perhaps due to extensive and intensive climatic changes such as galciation or due to rapid uplifting of the outer Himalayan chain.

Many of the pluri-zonal species in this vegetation type such as Achillee millefolium, Chrysanthemum leucanthemum, Sileneinflata, Verbascum thapsus have been considered as synanthropical immigrants. Some of the characteristic xerophillous shrubs of this type are regarded as outposts of sub-tropical semi-arid vegetation whereas other shrubs characterize by the contact with extra-tropical semi-arid flora of central Asia. This type is perhaps having the closest affinity to the arid and semi-arid vegetation of central Asia.

The Potwar plateau has an open cover of Pleacuspidata and the vegetation has been designated as sub-tropical evergreen sclerophuyllous forest while Kashmir valley has been considered as the eastern-most limit of the mediterranean climate in Asia. Here the Cedar foresters are not associated with Quercus baloot or Pinus gerardiana as in chitral, although in upper Chenab these species are encountered, while at lower elevations, Pinus roxburghii and Olea cuspidata, Punica grantatum are met with. In the Kashmir valley, plants like Poplars and Platanus orientialis, though introduced, are in harmony with mediterranean characters of the climate. Similarly mention may be made of Salix alba, S. babylonica, S. daphnoides, Morus alba, M. indica, and juglans regia, etc. Further east to the Kashmir valley,l Acacia, Carissa dry forest on foothills and 'colline' belt of west Himalaya has 55% tropical semi-arid elements. On rocky sites, Euphorbia royleana is a special variant of the thorn forests while under less extreme climatic conditions, Nerium indicum or Punica granatum occur in the sub-montane transition belt.

Olea ferruginea extends from subtropical dry forests to extra-tropical sclerophyllous forests. In Pinus roxburghii forest, Chir pine dominates where extra-tropical elements predominate over sub-tropical taxa. The evergreen and alaurophyllous species forming low scrub trees comprising 26 per cent of the taxa, are extra-tropical humid east-side elements. Both edaphic and biotic factors such as fire and high population density cause drier conditions in these forests with the result that tropical semi-arid and extra-tropical semi-arid species predominate.

Cedrus deodara and Pinus excelsa forests occur in the oak zone which also contains several extra-tropical humid species having relation to he east side humid flora. These conifer forests sometimes occur as "seral" in nature in the inner dry valleys because of the peculiar bio-edaphic conditions. Cedrus deodara itself extends south-eastwards up to east Nepa; in the zone of bixeric climate with both winter and summer precipitations. The species is much influenced by human activities. Vishnu-Mittre is of the opinion that Cedrus doedrus deodara has been derived from C. liboni, which migrated into the

Himalayas during Pleistocene. Studies on the fossil flora of Karewas in this connection are interesting. It is considered that the climatic conditions changing from glacial to inter-glacial have been responsible for the creation of new species or varieties in the Himalayas. This is further elucidated by the fact that none of the conifer is natural except Podocarpus nerifolius in the Peninsular India.

In the extreme western part of the Himalayas like Kashmir and inner dry valleys of Himachal Pradesh, laurophyullous species do not predominate and conifers grow along with the deciduous species like Acer, Aesculus, Fraxinus, Juglans, Populus, Prunus, etc. On dry sites, the shrubs like Parrotiopsis jacquemontiana are common. In spite of close floristic affinities between the Cedrus deodara forests of the mediterranean regions and of Western Himalaya, Variants of Juniperus and spiny cushion plants are regularly found in dry Cedrus forests of the mediterranean region and elements of evergreen laurophyllous and semi-evergreen deciduous trees are absent but are dominant eastwards. In their bio-coenological characters, Cedrus deodara forests show close relations to humid vegetation and least to semi-arid vegetation than to mediterranean cedar forests.

The Olea feeruginea-Quercus baloot communities of the inner and middle Chenab valley correspond to Quercetum ilex of the Mediterranean region. These forests in the Himalayas are dominated be plants like Daphne cashmiriana, Myrsine africana, Nerium indicum, Pyracantha crenulata, along with some sclerophyllous and trees and shrubs. During the dry season, some of hte plants remain leafless. In these forests, extra tropical elements are prevalent but many extra-tropical semi-arid species characterize the close affinity with the mediterranean vegetation. Nevertheless among these are several taxa that extned to semi-humid parts of east Asia.

Areas of only some taxa such as Fraxinus xanthoxyloides, Quercus ilex, Daphne oleoides are confined to Western Eurasia.

Certain exrophyllous shrubs with relations to sub-troical dry forests are also present. These grwo mainly in hot, sub-montane regions whereas Olea ferruginea dominates in the three zone. Phytogeographical grouping: (i) The mediterranean element coprises species of mediterranean region proper, those confined to east mediterranean region, species of extreme east mediterranean region and plants cultivated in mediterranean climate (ii) Mediterranean and temperate elements comprise species common to mediterranean region and temperate countries, species of temperate countries, species of temperate Asia, Sino-Himalayan species, Himalayan species, Himalayan species occurring also in east mediterranean region, as well as Himalayan species found extending east and those cultivated in mediterranean and temperate countries. Based on the distribution of taxa common to western Himalaya and mediterranean region, the following broad groups ave been made:

Actaea spicata, Arabis auriculata, Barbarea vulgaris, Buxus sempervirens, Campanula latifolia, Cardamine oxycarpa, Circaea luteriana, Cotinus coggyria, Fraxinus excelsior, Iris germanica, Lathrea squamaria, Nerium indicum, Phleum arenarium, Sanicula europaea.

(ii) Sub-mediterranean Cult-Spontaneous:

Caucalis latifolia, Erodium cicutarium, Scandis pecten-veneris, Verbena officinalis, Vicia angustifolia, V. tenuifolia.

(iii) Sub-mediterranean continental:

Bupleurum falcatum, Saxifraga granulata.

(iv) Mediterranean pontic:

Galium verum, Sambucus ebulis

(v) Euro-mediterranean:

Atropa belladonna

Tropical

(vi) Sub-mediterranean Irano-Torranean

Aleuropus littoralis, Alhagi maurorum, Anchusa italica, Apocynum venetum, Arthrocheum glacum, Bromus scoparius, Chenopodium murale, Chondrilla juncea, Elymus caputmaduse. Frankenia pulverulenta, Launaea nudicaulis, Linum corymbulosum. Lithospermum arvense, Matricaria aurea, Melica cupani, Nardus tenuiflourus, Silybum marianum.

(vii) Mediterranean Irano-Tourranean

Allium ampleoporaseum, Avena barbata, Capparis spinosa, Chrysopogon gryllus, Cressa cretica, Cynanchum acutum, Gageareticulata, Imperata arundinacea, Iris syrinohium, Lippia nodiflora, Silene colorta.

(viii) Mediterranean Irano-Tourranean and Saharo-Sindhian:

Erodium malacoides, Loegea purpurea, Stipa tortilis.

(ix) Euro-Siberian boreal american-mediterranean:

Aspergo procumbens, Euphorbia flacata, Galium setaceum, G. spurium, Melisa officinalis, Silene gallica.

Plants without mediterranean tendency: This group comprises the north-western dry elements growing in the plains of Indus, Saharo-Sindhian and Sudano-Rajasthan regions. These names have been given for the sake of reference although they do not extned to the Western Himalaya. As indicated by Aubreville, we can call it as Indo-Saharo-Sudanian also. However, recently Legris applied the term

Saharo-Sindhian and Sudano-Rajasthanian, respectively. Whatever term we may apply, all these western Indian and African areas testify liasion between the areas of these two countries, today separated by the sea and large deserts. The flora between these two regions is homologous constituting various vicariant species common to genera like Acacia, Anogeissus, Boswellia, Calotropis, Capparis, Carissa, Commiphora Flacourtia, Propsopis, etc.

Though the term vicariant is open to several interpretations, it connotes 'distinct endemic Kinnean species' as opposed to races which have been termed as micro-endemics'. Examples to the vicariance existing between the plants of the mediterranean region and of the Western Himalayas, are as given below. The woody plants of Western Himalayas show remarkable association with mediterranean vegetation. There are many closely related trees and shrub group in inner valley which are not influenced by monsoon winds but are open to westward drift in winter. Few examples of such elements showing vicariance are as follows:

Mediterranean	West Himalayan
Atropa belladonna	A. acuminata
Berberis hispanica	B. aristata
Brachypodium sylvaticum	B. sylvaticum
Calamintha clinopodium	C. clinopodium
Carduus ballii	C. edelbergii
Cedrus atlantica	C. deodara
Cotinus coggygria	C. coggygria
Cotoneaster racemiflora	C. roseus
Daphne oleoides	D. cashmeriana
Geum urbanum	G. urbanum
Hedera helix	H. nepalensis
Organum vulgare	O. vulgare
Quercus ilex	Q. baloot
Viburnum lantana	V. cotinifolum

Among the conifers, the genus *Abies*, having a number or species in the mediterranean region and in the Himalayas, is separated as webbiana, pindrow, gambli *and* brevifolia.

Pinus, a complex genus has a number of mediterranean species. for instance, *P. sylverstris.* a largely distributed species in Eurasia is not represented in the Himalayas but among the 5-leaved forms it is represented in the Himalayas by *peuce, excelsa* and dilatensis and from the group *gerardiana* by *P. longifolia.*

Cupressus, a plant of the east mediterranean regions with sempervirwna is represented till Kashmir by an endemic pykashmiriana and *torulosa* in other parts of the Himalaya.

Cedrus, a mediterranean genus is represented with *deodara* in the Himalayas.

Juniperus existing in Europe before glaciation as *nana* may be a form of *communis; sabina* is found up to central Asia, *excelsa* in rsia, and *macropoda* from Caucasus to Nepal.

From Palmacceae, *Chaemorops maritima* of hte Himalayas corresponds to *C. humilis* of West mediterranean areas.

From Fagaceae, among the numerous species of *Quercus, Q. ilex* attains its limit till the Bashahar Himalaya.

Celtis, a mediterranean genus is represented by *caucasica.*

Parrotiopsis jacquemonitiana present in the Himalaya is represented as *persica* in the West mediterranean region.

Olea cuspidata of the Himalayas corresponds to *O. europea of the mediterranean region*

Thesium himalensispy of the Himalayas corresponds to T. humile of the mediterranean region.

Distribution of a few common species is described as under:

Rosaceae represented in India has two centres of origin: (i) some genera originated from China while others like Prunus, Rubus, Rosa, Cotoneaster and Purus in W. Asia. Plants migrated in both the centres by West and East route by way of Afghanistan, Kashmir and North Burma. (ii) Rosa moschata of the Himalaya shows affinity with R. sempervirens of mediterranean region.

Ilex aquifolium in middle europe and mediterranean region has many species in the Himalayas.

In Lamiaceae, Nepeta, Origanum, Phlomiz, Salvia, Stachys, and Teucreum had their origin in mediterranean region and found their way to India through west.

Quercus ilex, a small evergreen tree or large shrub in the inner dry valleys of the Himalaya is common in Kunawar, the upper Ravi valley and Kagan; not in the outer Trans-Indus hills. Usually, gregarious and associated with Acer pentapomicum are Fraxinus xanthoxyloides and Pinus gerardiana with Plectranthus rugosus. The tree is injured by cattle browsing and the leaves on low branches become pricklyl-toothed much like those of European Holly. The wood is hardly used.

Fraxinus excelsior is wild in Hazara, Kashmir and Chamba hills of Himachal Pradesh. It differs from *F. excelsior* of Europe in having fewer leaflets and longer and more slender racemes and is specifically considered distinct as *F. hookeri.*

Nerium indicum is found on the beds of streams in the Himalaya ascending to 1,900m and in sub-Himalayan tracts, does no extend far into the Punjab plains.

Hippophae rhamnoides L. is a very variable plant from Kunawar to Kumaon. One of the variety with scaly twigs and distinct scales on youngest shoot only, soon forming a silvery-waxy covering of the shoot is like the Sea Buck thorn of Europe and is present in the inner dry valley of the Himalayas from 2,700-3,700m. Another species, *H. salicifolia* distributed from Kunawar to Kumaon along the stream banks of inner Himalayas is somewhat larger than *H. rhamnoides* but not so common in drier parts where it is replaced by latter.

Buxus wallichiana is a small tree found gregarious in patches usually in moist shady places in Chamba, Kangra, Kulu, Simla and Bashahar up to Kumaon.

There is no agreement among the various authors on the phytogeographic limit and subdivisions of the mediterranean region. The main diversity of opinion is: (i) that whether the mediterranean region is a natural circumscribed region or is a part of the vast region that embraces the oriental countries till Afghanistan and Baluchistan; (ii) that the steppic region of Grisebach is a natural region, if not, are these the natural regions circumscribed; (iii) the flora of physically naturally isolated or common with Arab-Sind. What are the relations of the flora of Africa with Sahara and (iv) are there special affinities between these two regions.

Indicated Indo-Pakistan sub-continent to have three floral regions: (i) Indo-African including peninsular India, Gangetic plain, Assam and Burma; (ii) Mediterranean including Rajasthan valley of Indus and Kashmir and (iii) the flora of Sahara, a part of palaeotropic flora forms a region phytogeogra-phically naturally isolated or common with Arab-sind. What are the relations of the flora of Africa with Sahara and (iv) are there special affinities between these two regions. Gaussen indicated Indo-Pakistan sub-continent to have three floral regions:

(i) Indo-African including peninsular India Gangetic plain, Assam and Burma; (ii) Mediterranean including Rajasthan valley of Indus and Kashmir and (iii) Holarctic including Himalayas. In a broad basis, Good considered India into two floral empires: Boreal covering Sino-Japanese regions or province of Sino-Himalayan-Tibetan mountains and palaeotropical Kingdom, having 2 sub-kingdoms, the African comprising desertic regions of north Africa and India and Indo-Malayan. Clarke and Hooker were, however, the first to define floristic divisions for the country. However, it is evident that Himalayas on the side and desertic regions of the Indo-Pakistan sub-continent on te other, imposed barrier for separating floral regions.

The region of Assam and Bengal has strong phytogeographical afinities with the Indo-Malayan region. Hooker grouped species of the Indian flora into types such as European, Siberian, and Malaysian and emphasized the role of Indian Continent as a cross road to different types of floras. The view was later supported by Carke and others till Chatterjee showed it to be only partly true considering Indo-Gangetic plains where there is little to suggest indigenous vegetation. Though there are connections in the north and west, India has her own flora with 28.8 per cent of endemic elements in the Himalayas.

Distribution of flora is subjected to two principal types of climate: (i) Dry desert belt from Sahara across Arabian Peninsula and Iraq and (ii) A sub-tropical element from Turkey, Persia, Iraq and Afghanistan. Chatterjee showed that 11% of the Indian flora with certain families constitute their best development in the mediterranean region which may be regarded as their original home. From here onwards, a natural process of migration has taken place both east and west and the number of migrant are thinner and lesser and directly vary with the distance from the main source. Broadly, it could be said that Boraginaceae, Brassicaceae, Capparidaceae, Caryo-pyllaceae, Fumariaceae, Lamiaceae and Rosaceae, have their origin in the orient.

Whether the 'orient' interpreted by Chatterjee as mediterranean should be taken up as a separate floral region remains to be answered. Brassicaceae as a whole, was regarded as mighrant from east mediterranean region. Some of the species from orient to India are: Brassica tournefortii, Capsella bursa-pastoris, Cardamine impatients, Europhila vulgaris, Erysium repaudum, Farsetia aegyptiana, F. jacquemontii, Lepidium latifolium, Malcolmia africana, Mathiola odoratissima, Moricandra arvensis, Nasturtium officinale, Nestia panticulata, Sisymbrium irio. S. loesellii, S. sophia, S. wallichii, etc. Based on the available literature, a list of the available plants of Indo-Pakistan subcontinent in countries towards east and west of the mediterranean sea was drawn which show the following results:

Total number of families represented in the region east and west of mediterranean sea = 66

Total number of common plants = 377

Plants represented in different countries:

1. Mediterranean countries comprising Steppico-Desertic regions: (a) Spain, France Medit, Ital Medit., Yugoslavia and Albania Medit., Greece Crimee, Asia minor, Chypr.e, Syria, Sinai, Egypt, Cyrenae, Tripolitaine, tunisia, Alteria, Moracco, Canary islands.
2. Oriental countries comprising of southern Russia-Extra-tropical Arab, Mesopotamia, Persia. Afghanistan, Turkistan, Aralo-caspian, Trans-Caucasus, S. Russia, Palestine-presently Israel.

3. Tropical steppic region: Tropical Arab, Nubia, Abyssinia-Sudan-presently Ethiopia.
5. Tropical and South Africa, Tropical and east Asia, Australia, South and Central America.

Among the salient conclusions which could be drawn from the present study are:

1. The eastern limit of hte mediterranean climate can roughly be fixed on the west of river Indus having encalves in the Kashmir valley. An evaluation of the detailed climatic maps of each valley would certainly reveal the extent of hte mediterranean influence in this region.

   The distribution of the mediterranean flora was subjected principally to two types of climates: (i) To the dry desert belt from Sahara across the Arabian Peninsula and Iraq and (ii) along the subtropical climatic belt running from Turkey, Persia to Afghanistan.
2. In the sub-tropical formations on the outer chains of the Himalayas, Pinus roxburghii in the Central and Pinus gerardiana in the western part of the Himalayas form a transition with the mediterranean formations. In the he temperate zone where the range of rainfall is variable, depending on the aspect and position of the hills and hte temperature of the coldest month is not less than 10°C, Quercus incana in central and Q. afflex in the western part of the Himalayas form of transition with the mediterranean formations.
3. The region of Kashmir forms a transition with the mediterranean and monsoonic region of the Himalayas and forms the floral limit of the Mediterranean elements. However, the distribution of certain species in the regions east of Kashmir, discussed in the article, show enclaves which are

protected from the influence of monsoon and resemble very much within the mediterranean climate.

4. Similarly, analysis of the north-west Indian flora by Meher-Homji has shown that in the areas experiencing tropical climate percentage of the typical mediterranean elements is very low as against 89.0-96.5 per cent of tropical elements.
5. The winter rains which occur occasionally may account for low percentage of typical mediterranean elements. for bixeric climate with winter spring rain typical mediterranean element is 7.1-13.7%. In areas having moderate degree of mediterraneity like Baluchistan, typical mediterranean element is as high as 25.5-40.4%.

Floristic surveys of different regions are extremely essential in order to furnish complete understanding of vegetation of the past and present, as well as to support various types of studies in botany and allied modern disciplines. With these objectives in view the authors selected one of the important regions of Garhwal Himalaya for an enumeration of the plant wealth.

Gopeshwar happens to one of the newly developed township of the region and experiences different types of biotic pressures on plant life which are specially accelerating with the process of urbanisation. Therefore, extensive field surveys were carried out for two years in Gopeshwar and adjacent areas. The communication lists 691 species under 417 genera and 109 families. Chamoli Garhwal situated in the central part of the Western Himalaya, constitutes inseparable part of Indian culture and mythology since the dawn of history and even prehistory. It has attracted people with varied taste and objectives for its beautiful mountains, valleys, rivers, flora, fauna, snowy peaks, religious shrines like

Badrinath, Kedarnath, etc. Plant exploration in Garhwal dates back to 1796 when Thomas Hardwicke started collecting from the Alkananda valley, followed by W. Moorcraft while he was crossing through Niti pass.

Comprehensive collections were made during the year 1846-1849 by Strachey and Winterbottom. Their account was revised by Duthie, who also included the work of Royle, Wallich, Falconer, Blinkworth, Thomson, Edgeworth, and others. Osmaston after his 13 years of collection published "A Forest Flora Of Kumaon", which included shrubs and trees while trekking from Srinagar to "Valley of Flowers" prepared a list of plants. Rau also made comprehensive collections from the interior parts of North Garhwal followed by Naithani and Singh & Singh who explored the adjacent localities of Garhwal Himalaya. A perusal of literature indicates that earlier explores paid little attention to this zone of Himalaya. The present exploration was carried out with a view to compile a list of plants, as complete as possible, that exist at Gopeshwar and nearby places. The study would be helpful to the scholars working on various aspects of vegetation from this interior part of the Himalayan basin.

The area under study includes Gopeshwar and adjacent places of Dasholi Block, ranging from 700 to 4,200m of altitude, lying in the district of Chamoli Garhwal. The town of Gopeshwar is situated at an altitude of 1,300 to 1,500m on intermediate hilly ridges which continue higher towards North East and lie between 30°N latitude and 79—25′E longitude. It is encircled by high hilly ranges followed with slopes and the valleys. The area is delimited by river Alaknanda and its tributory Balkhila in the South East and North West, respectively.

All villages and the town of Chamoli, comprising the area under study, are situated up to 2,000m above sea level. In the north of Gopeshwar there are famous shrines and temples of Tunganath, Ansuya, Chandrashila Peak and villages of Mandal, Bairagana, Sagar, Gangolagaon, while religious shrines of Rudranath and village namely Rauli, Dewar, Khadora are on its North East.

Snowy peaks of Nanda Ghunghati are visible marginally being situated far away after some hilly ranges in the East. Along the banks of Alaknada river in the South there are villages such as Paduli, Kothiyalasin, Chamnau, Patiyaldhar, Balkhila, Maithana and a small township, named Chamoli (formerly known as Lalsanga) which is on the main road leading to Badrinath. In the West there are Gair, Tangsa and Bamiyala villages of the Balkhila valley. With the lapse of time, changing course of water streams, lanslides, floods, vegetational degradation, other biotic factors and electric thundering, especially at higher elevations, have caused variations in general topography of mountains and vegetation. Valleys with cliffs and gorges have all sort of exposure and faces.

Basic types of rocks present in this area re granite, dolerite, epidiorite, and igneous. Sedimentary rocks consist of limestone and stone, conglomerate, breccia, shale, etc. wheel quartzite, dolomite, shlorite and mic aschists are abundant in the metamorphic rocks. Soil is mostly sandy along the back of Alkananda and to some extent in other ravine valleys. It varies from alkaline sandy loam to clayey loam in open agricultural fields of low lying areas while there is humus rich soil in oak forest, acidic in pine forest and gravel to chert type in alpine region. Alkalinity, acidity and mineral nutrients also vary from place to place depending upon the nature of parent rock and its units.

Several ecological factors are responsible for a unique feature of the climate, varying from sub-tropical to alpine with the usual rainy winter and summer seasons. Frequent rains occur at high elevations but the total amount of precipitation decreases with increasing elevation. Maximum rainfall occurs during rainy months of July and August. Heavy and early snowfall along with frequent rains are common in alpine zone with gradual decrease and late occurrence common in alpine zone with gradual decrease and late occurrence down to 1, 500 m of altitude during winter months of November to February. Temperature in valleys during this period ranging from 15° to 24°C, comes down to 0°C while there is a gradual upward trend from March onward and it reaches maximum during late May to early June.

Corresponding to the climate and topography of the area, there are sub-tropical, temperate and alpine types of vegetation ranging from 750-1, 600m, 1,600-2,900m and 2,900-4,200m of altitude, respectively. Different factors have influenced the normal limits of vegetation types and flowering seasons. Natural vegetation near towns and villages in the sub-tropical belt has been subjected to more biotic disturbances as compared to the villages at higher elevations. There is abundance of grasses and annuals during monsoon, whereas perennials, shrubs and trees mostly bloom during spring and summer.

Sub-tropical zone comprises mainly xerophytic vegetation throughout the bank of Alakanandas and in the lower portion of Balkhila valley. It is represented by shrubs like Ziziphus mauritiana, Caesalpinia decapetala, Euphorbia royleana, xanthium strumarium, Opuntia vulgaris, Woodflordia fruticosa, Cryptolepid buchananii, Artemisis scoparia, Carissa spinarum, Adhatoda vasica and trees e.g. Mallotus philippinensis, Sapium insigne, Emblica officinalis, Rhammus triqueter, Bombax ceiba, Boehmeria rugulosa, and Phonex humilis.

Deep in the Balkhila valley, this vegetation is replaced by the plants of ravines and include trees of

Alnus nepalensis, Spondias pinnata, Engelhardtia spicata, Pyrus pashia, etc. and characterostoc shrub such as Callicarpa macrophylla. Other shrubs are Plectranthus japonicus, Craniotome versicolor, Buddleja neemda, Rhamnus triqueter, Urtica dioica, Arundo donax, Artemisia nilagirica, Berberis asiatica, Rubus ellipticus, and Rosa brunonii. Some of them are also common along road side and on waste lands, and include climbers such as Galium, Clematis, Ipomoea, Dioscorea, Cissampelos, Stephania, and Rubia.

Some of the common herbs and weeds are Rumex hastatus, Dicliptera roxburghiana, Reinwardtia indica, Coleus forskohlii, Tridax procumbens, Cynoglossum spp., Strobilanthes spp., Polygala abyssinica, Crotalaria calycina, Seigesbeckia orientalis, Micromeria biflora, Thalictrum foliolosum, Boenninghausenia albiflora, common weeds Cannabis satica, Tagetes minuta, Chenopodium ambrosioides, Cirsium wallichii, Erigeron spp., Euphorbia spp., Euphorbia spp., an Bidens spp. etc. On steep metamorphic slopes, opposite to Alakananda and Balkhila and deep in the North of Gopeshwar, there are forests of Pinus roxburghii ranging from 1, 000-1, 600m of altitude. Undergrowth of such foreses mostly consists of grasses and occasional shrubs.

On other aspects of the slopes starting from valleys, grasses such as Arundinella nepalensis, Apluda mutica, Andropogon tristis, Cymbopogon distans, C. martinii, Oplismenus spp. Agrostis, Eragrostis, Bothriochloa, Panicum, Cynodon dactylon, Saccharum spontaneum and Cyperaceous members e.g. Carex condensata, Eriophrum comosum etc. constitute the grassy slopes. Associated herbs are species of Desmodium, Crotalaria, Galium Cassia, Swertia, Bupleurum, Phaseolus, Vicia, Striga, and others. Wider portion of the valleys and terraces of surroundings slopes of mountains are useful from physio-agronomic points of view. Crops are grown twice a year besides seasonal vegetables, fruits etc. Important crops are of major cereals-wheat, paddy, barley, maize, millets, finger millet, Setaria italica, Panicum frumentaceum pulses-Pisum sativum, Phaseolus mungo, Glycine max, Lens culinaris, Cajanus cajan and oil crops like Brassica compestris, Sesamum indicum etc. Common weeds of winter season crops, which are locally known as Kuru, include species like Anagallis arvensis, Zornia gibbosa, Lathyrus aphaca, Fumaria indica, Stellaria media, Phalaris minor, Vicia hirsuta, Antirhinum orontium, Soliva anthemifolia, etc. Among weeds of rainy season crops, which are collectively known as Guderu or Challa, are Cyperus spp., Polygonum spp., Euphorbia spp., Chenopodium spp., Galinsoga parviflora, Oxalis corniculata, Panicum psilopodium Fimbristylis dichotoma, Echinochloa crusgalli, and others.

Near villages, trees are raised for fruits, fodder and other purposes e.g. Celtis caucasica, Grewia optiva, Juglans regia, Citrus spp., Toona serrata, Ficus spp., with occasional association fo Mangifera indica, Carica papaya, Ficus religiosa, F. bengalensis in the lower limits of the valley. Farms of sericulture include the cultivation of Morus alba and Quercus leucotrichophora to yield raw silk and tussar, respectively.

Many ornamental an shady trees have been introduced on the oradsides e.g. jacranda mimosaefolia, Bauhinia variegata, Cassia fistula, Eucalyptus spp., and a few others. With the end of upper limits of pine forests and ont he northern slopes above Gopeshwar, there starts temperate mixed vegetation. Forests are locally known as Patal. Major tree components are of Quercus leucotrichophora associated with Myrica esculenta, Rhododendron arboreum, Litsea umbrosa, Alnus nepalensis, Lyonia ovalifolia, Osmanthus, fragrans, Cotoneaster bacillaris, etc. Ground vegetation, in exposed forest pathces, consists of shrubby species e.g. Berberis chitria, Deutzia corymbosa, Elaeagnus umbellata, Pyracantha crenulata, Girardinia palmata, Princepid utilis, Rubus lensis. Among the common herbs Pedicularis spp., Plumbago zeylanica, Brunella vulgaris, Geranium spp., Primula denticulata, Gentiana spp., Potentilla spp., Myriactis nepalensis, Swertia chirata, Anaphalis margaritacea, Trifolium repens, Bergenia ciliata, Circium arvense and grasses like Erianthus rufipilus and Miscanthus nepalensis constitute the ground vegetation.

Shady cool aspects are characterised by trees like Quercus florilbunda, BEtula alnoides, Carpinus viminea, Eurya acuminata, shrubs i.e. Daphne papyracea, Sarcococa saligna, and Arundinaria spp. grasses Chrysopogon gryllus and Eulalia quadrinervis; andcyuperaceous species and herbs lije Taraxacum offiicinale, Tragopogon gracile, Habernaria pectinata, Salvia lanata, Androsace lanuginosa, Erigeron spp., and several others. The higher elevation limits of temperate vegetation are represented by trees like Quercus semicarpifolia, Syringa emodi, Abies pindrow, Texus wallichiana, Acer caesium, Betula utilis and the ground vegetation includes Leycesteria formosa, Salvia glutinosa, Polygonum spp., Ophiopogon intermedius, etc.

Alpine vegetation is usually of herbaceous nature, locally known as Byghals with procumbent bushy patches of Rhododendron barbatum, etc. at lower extremities. Herbs with their limited span of beautiful flowering being to sprout as snow starts melting. They include species of Polygonum, Primula, Aconitum, Potentilla, Cynanthus, Orchis, Morina, Valeriana, Anaphalis, Inula, Saussurea, Senecio spp. Allium, Gentiana, Ranunculus, Thalictrum, Cotoneaster and abundance of soft grasses and ferns like Osmunda spp. Flowers of Saussurea obvallata find a place in the temples of Rudranath as being offered during worship.

Grazing is common during rainy season by sheep, goats, mules, buffalowes of Gujars and local inhabitants. While the genera and species are arranged alphabetically, the families are listed according to Bentham and Hooker's system of classification with slight modification as per modern concept.

An attempt has been made to incorporate the latest change sin botanical nomenclature of plants, Local/ vernacular names of some plants have been given within brackets following the botanical names, after verification from the local inhabitants to help in the future identification. Each species is followed by a brief not about its habit, colour of flower, nature of inflorescence, peak flowering time, habitat, approximate altitude or range of occurrence and the corresponding herbarium number.

Kumaon Himalaya is perhaps the most attractive regions from the scenic as well as the floristic point of view among all the sectors of the Western Himalayas. It is located between 28° 43° to 30° 49°N latitudes and $78^{0}44'$ to $81^{0}4'$ E longitudes and includes three districts viz. Naini Tal, Almora, and Pithoragarh. The high altitude regions are chiefly in the district Pithoragarh especially in Dharchula and Munsiari tehsils except Pindari which fall sin Almora district. The important places from where plants have been collected earlier are Pindari and its morainic environs, Panchuli basin, Milam and Shelong Glaciers, Suryakund, Sangcha kind, Trisul basin, Brijikang pass, Ralam valley, Upper Dwali and Kali valley. Several other places have been botanised by the author during the past five years.

The prominent plants forming the general vegetation of alpine region are : Betula utilis D. Don, Rhododendron campanulatum D.Don., R.lepidotum Wall. ex G. Don., R. anthopogon D. Don.Lonicera angustifolia Wall. ex DC., L.

obovalata Royle ex hook.f. & Thoms., L. purpurasens Hook.f. & Thoms., L. webbiana Wall. ex Lindl., C. acuminatus Lindl., Ribes glaciale Wall., R. emodense Rehder, Berberis umberllata Wall. ex G. don., B.kumanonensis Schneid., Rosa sericea Lindl., R. macrophylla Lindl., Caranga gerardiana Royle., C. pygmaea D., Hippophae rhamnoides Linn., Myricaria elegans Royle, M.germanica Desv., Salix fruticulosa Anderss., S. sclerophylla Andress., S. wallichiana Anderss., and several others. These are conspicuous woody representatives of the alpine flora and the herbaceous elements are represented by species of Astragalus, Corydalis, Polygonus, potentilla, Saxifraga, Gentiana, Impatients, Sedum, Pedicularis, Senecio, Primula, Epilobium,/saussurea, Swertia, Rumex, sedges and grasses. Plants have been seen to grow generally up to 5,600m.

The alpine zone usually begins where the tree zone ends. This is known as the timber line. A fringe of shrubs, scattered or continuous dwarf or gnarled trees, is found immediately above the tree line. These stunted trees exhibit a characteristic wind-wept or snow-wept appearance because of their exposure to the fury of cold winds, blizzards and snow storms. Birches, rhododendrons and willows among others are usually seen in such places. The high altitude conifers, e.g. species of Abies, Pecea, Cedrus, and Pinus are seen associated with broad-leaved trees like Quercus semicarpifolia and Betula utilis at the altitude of about 3,600m. B. utilis may often be seen forming pure stands or usually forms association with Rhododendron campanulatum.

Among the shrubs found above the tree line are usually species of Berberis, Myricaria. Lonicera, Rhododendron, Salix, and Jumpierus. While going on excursion in high altitudes, one usually finds a grassy belt with lush meadows, grassy slopes and flower-laden pastures along with several prostrate shrubs like willows and heathers above the tree line. These meadows and grassy slopes offer an important place for alpine plants and are well-known for their beautiful and attractive flowers. In late May and early June, the summer flowering herbs being to flower in great profusion. Among them are several crucifers, primulas, anemones and saxifragas and late in summer and towards autumn appear the gentians, swertias and others with saussureas attaining their best development late in autumn. In Kumaon, the high altitude pastures and grassy slopes are locally called as bugyals. These are generally found at altitudes above 3, 500m and several beautiful and graceful herbs occur in great profusion on the gentle slopes in the bugyals.

The bugyals generally show mixed association of grasses, sedges and colourful herbs, belonging to various families but sometimes a pure association of grasses is also seen. These grassy meadows from a lush carpet in many localities particularly in Upper Gori valley. The distribution pattern of bugyal flora is largely governed by the local edaphic and climatic conditions. The varied, abundant and colourful high altitude flora of Kumaon Himalaya has been explored by several professional as well as amateur enthusiasts. Among the persons who had involved themselves in this great work include administrators, army officers, missionaries, surgeons, and survey officers who contributed towards most for the advancement of present day knowledge of flora. William Moorcraft, Trail, Lindsay, Edward Madden were the collectors in this area during the early years.

The most extensive and the best collections in Kumaon was that of two survey officers, Strachey and Winterbottom. A list of which was later published by Duthie. George King, Duthie and his collectors, Wallich,

Falconer, Royle, Drummond, etc. explored this area towards the end of the past century. Two important publications on the orchids of North Western Himalaya and a revised list of plants of Kumaon and Garhwal, were authored by Duthie. Osmaston brought out his 'A Forest Flora for Kumaon'. He described and enumerated only trees, shrubs and climbers but did not include herbaceous vegetation. Several other workers have studied the Apline flora of Kumaon.

Analysis of present day alpine flora of Kumaon Himalaya reveals 662 species belonging to 251 genera and 57 families. The number of genera and species have been given in parentheses for each family. The families represented in this zone are: Compositae (27/80), Ranunculaceae (16/39), Cruciferae (23/37), Rosaceae (9/36), Papilionaceae (12/34), Gramineae (18/32), Saxifragaceae (4/30), Caryophyllaceae (9/29), Scrophulariaceae (7/28), Primulaceae (3.22), Cyperaceae (4/21), Umberlliferae (12/20), Gentianaceae (4/19), Polygonaceae (4/19), Liliaceae (12/18), Lamiacace (10/16), Fumariaceae (1/14), Caprifoliaceae (3/14), Crassulaceae (2/13), Juncaceae (2/12), Onagraceae (1/10), Orchidaceae (5/9), Campanulaceae (3/9), Ericaceae (3/6), Berberidaceae (2/5), Valerianaceae (2/5), Papaveraceae (2/4), Geraniaceae (1/4), Balsaminaceae (1/4), Rubiaceae (1/4), Oleaceae (2/3), Solanaceae (3/3), Dipsacaceae (3/3), Orobanchaceae (1/3), Urticaceae (2/3), Violaceae (1/2), Tamaricaceae (1/2). However families like Araliaceae, Betulaceae, Cannabinaceae, Convolvulaceae, Fagaceae, Iridaceae, Circaeasteraceae, Asclepiadaceae, Lentibulariaceae, Loranthaceae, Polemoniaceae, Potamogetonaceae, and Zannichelliaceae merely have only 1 genus and 1 species.

Some interesting plans of these regions are: Aconitum heterophyllum Wall. ex Royle, A. violaceum Jacq. ex Stapf., Adonis chrysocyathus Hook f. & Thoms.,

Allium carolinianum DC., And rosace globifera Duby, Anemone obtusiloba D. Don., Aster diplostephioides C.B. Clarke, Bergenia strachyei Engl., Clematis barbellata Edgew., C. orientalis Linn., Corydalis cashmeriana Royle, Cyanathus lobatus Wall. ex Benth., Delphinium cashmeriana Royle, D. pyramidale Royle, Epilobium latifolium Linn., Gaultheria trichophylla Royle, Geranium pratense Linn., Hippophae tibetana Schlect., Inula grandiflora Willd., Iris kumatogonium carinthiacum Reichb., Meconopsis aculeata Royle, M. robusta Hook. F & Thoms., Megacarpaea polyandra Benth., Onosma hispidum Wall. ex D. Don., Orchis latifolia Linn., Paraguilegia anemonoides Ulbi., Pedicularis pectinata Wall., Phlomis bracteosa Royle ex Benth., Picrorhiza kurrooa Royle, Pinguicula alpina Linn., Pleurospermum candollei C.B. clarke, Polygonum affine d. Don., Potentilla atrosanguinea Lodd., P. fulgens Wall. ex Hook. f & Thoms., Primula macrophylla D. Don. P. reidii Duthie, Saussurea gossypiphora D. Don., S. obvallta Sch-Bip., Saxifraga odontophylla Hook., F & Thoms., Scutellaria prostrata Jacq., Sedum wallichianum Hook., Senecio graciliflorus DC., Thermopsis barbata Royle, Trollius acaulis Lindl., Waldheimia tomentosa Regel, and several others.

Some of the plants like Thlaspi andersonii, Chorispora sabulosa, Christolea himalayensis, Creamanthodium nanum and others have a restricted distribution being found only in Kumaon Himalaya and in Western Himalaya or in adjacent Tibet. There are species like Thalictrum alpinum, Ranunculus hyperboreus, Potentilla multifida, Oxyria didyma and several others which enjoy a world-wide distribution in the alpine as well as arctic localities.

Cicer soongaricum and Lamium rhomboideum extend north wards to the Central Asian high-lands. Gueldenstaedtia himalaica, Saussurea leontodontoides,

Cicerbita macrorhiza, Picrorhiza kurrooa along with others are distributed along the Himalaya from west to east. Alectris pauciflora and Anemone repicola are found in West China and also throughout himalayas. Senecio coronopifolius with a few others are distributed westwards to Afghanishtan, Iran and Caucasus.

In the sub-Himalayan tracts of Garhwal region, during the summer months, there are frequent man-made and natural fires. Their persistent and prolonged effect deflects the natural vegetation inducing the development of grasslands with stray trees. Several of the plants which appear in such grasslands are perennial herbs only a few centimeters high but develop below the ground enormous water storage tissues in the from of stout root-stock almost looking like the trunk of a small tree. A survey conducted in this region has shown that the plants which are able to survive through fires are quite frequent. A few of these have been described in the present communication Forest fire plays an important role in the development of vegetation.

In Africa considerable research has been conducted on the prolonged effect of fibre on vegetation and it was shown that many a fire resistant plants exist. In India, the situation has been reviewed recently by Some Deva *et al.* In the sub-Himalayan tracts of Garhwal and Dehra Dun during the summer months, there are frequent forest fires and their persistent and prolonged effect deflects the natural vegetation inducting the development of grasslands with stray trees.

Many of the plants which later appear in such grasslands are perennial herbs, just a few centimetres high with stout root-stock almost, looking like the trunk of a small tree and with enormous water storage tissue. The fire kills the aerial parts but the underground portion is little affected. As such soon after the fire, in favourable conditions, it sprouts to give rise to new shoot(s) and later conspicuous large blossoms, produce fruits, seeds and on the onset of rains develop small herbaceous vegetative branches. Surprisingly, the nearest relative of these plants are usually big trees or shrubs and further confirms that constant action of fire over long periods have resulted in the production of different plants which may finally lead to the creation of new species. Brandis was the first to point out the existence of this habit in Indian plants.

A small undershrub with a thick woody root-stock which produces annually after the fire of the hot season, herbaceous shoots with cuneate dentate leaves about 18 cm long and beautiful large purple flowers. It is found in the open grasslands on the Siwalik hills, south of Dehra Dun district and south Pauri Garhwal. The flowering times is normally between March and April but when the vegetation around is burnt, these could be seen as late as March to June and as early as January. The nearest relative of this species is C. *arborea* which is a large forest tree and in all respect except the size resembles the vegetative and floral features of the taxon under discussion

A small undershrub with a thick woody prostrate of underground stem producing annually 15-45 cm long branches. The flowers are small, produced in long, simple racemes. The fruit is around 4 cm long, four-winged and of beautiful pink hue. The mature leaves appear on the plant after flowering. It is rather common in burnt jungles, grasslands and on fire lines in Dehra Dun, Siwaliks, south Pauri, Garhwal, and outer Himalayas. Most of the other closely related species of Combretum are generally large, climbing shrubs.

A remarkable undershrub with a short, woody underground stem throwing up annually after the forest fires a number of herbaceous shoots 35-70 cm high. The flowers are yellow and appear in three-flowered umbels. The fruit is slightly lobed and its seeds are enclosed in sweetish pulp which is eaten quite often. It is rather common in grasslands in the region of Siwaliks, Dehra Dun, south Pauri Garhwal, and outer Himalayan ranges. It is characteristic plant of the tracts where forest fires occur periodically.

A dwarf undershrub up to 15 cm in height with a very short, brown tomentose stem springing annually from woody root-stock. The leaves are very large with long, winged petioles and congested racemes of small reddish flowers, appearing before the leaves. The plant was originally discovered from the banks of Ganges, near Haridwar by Colonel Hardwicks who came to Srinagar on a politically motivated mission to the King of Garhwal. We have been able to collect his plant once again from the type locality.

A low, deciduous undershrub with think, woody root-stock, annually producing numerous diffuse branches. Leaflets range from 3-5 and rarely up to 9, produced in long racemes, directly from the woody root-stock, before the appearance of leaves. It generally occurs in open grasslands in Dehra Dun, Siwaliks, and south Pauri Garhwal. This remarkable plant was first discovered by Buchanan-Hamilton in the year 1814 in and around Gorakhpur district.

A low suberect undershrub, rarely climbing; branches rustypuberlous. The leaves are coriaceous, ovate to lanceolate, flowers appear in corymbosely branched cymes. The plant has been seen in south Garhwal rather commonly. We have, however, collected it recently form the fire tracts of Bibiwala forest, Dehra Dun. It was originally reported in 1810 by Buchanan-Hamilton from north Oudh. Usually the growth of the primary stem is arrested and it develops into a thick woody root-stock from which annual stems are produced which usually flower during early summer, but may get destroyed before the fruit ripens by periodic fire.

Olex nana Wall. (Olacaceae): As small undershrub with a contracted woody stock from which herbaceous rod-like shoots are produced annually during the rainy season. Leaves sessile, lanceolate glaucous with recurved margins. Flowers small, white in colour, one each in axil of leaves. It is found in grasslands on the crest of Siwaliks and dry outer hills of Garhwal.

A small almost stemless undershrub with a flat rosette of 2 to 3 pairs of sessile cuneate pubescent leaves in false whorls, because the internodes are very short. Flowers are white in dense corymbs and the drupe is black globose. It is found in open grounds on the fire lines and the forests prone to annual fires. It is a very good example where due to the constant effect of the fire the whole plant has become so stunted that only few pairs of leaves are produced, while the other species in the same genera are usually big shrubs and trees.

A stunted 30 cm to 1 m high shrub with a subterrain root stock from which annual shoots are produced after the forest fire. Leaves orbicular, silky, flowers in 3-10 flowered panicles. It is found in open grassland and on the edges of sal forests. This controversial species is being added for the first time in the list of fire resistant plants which have become stunted, while their closest relatives are huge lianas.

Floristic surveys of different regions are extremely essential in order to furnish complete understanding of vegetation of the past and present, as well as to support various types of studies in botany and allied modern disciplines. With these objectives in view the authors selected one of the important regions of Garhwal

Himalaya for an enumeration of the plant wealth. Gopeshwar happens to one of the newly developed township of the region and experiences different types of biotic pressures on plant life which are specially accelerating with the process of urbanisation. Therefore, extensive field surveys were carried out for two years in Gopeshwar and adjacent areas. The communication lists 691 species under 417 genera and 109 families. Chamoli Garhwal situated in the central part of the Western Himalaya, constitutes inseparable part of Indian culture and mythology since the dawn of history and even prehistory. It has attracted people with varied taste and objectives for its beautiful mountains, valleys, rivers, flora, fauna, snowy peaks, religious shrines like Badrinath, Kedarnath, etc. Plant exploration in Garhwal dates back to 1796 when Thomas Hardwicke started collecting from the Alkananda valley, followed by W. Moorcraft while he was crossing through Niti pass.

Comprehensive collections were made during the year 1846-1849 by Strachey and Winterbottom. Their account was revised by Duthie, who also included the work of Royle, Wallich, Falconer, Blinkworth, Thomson, Edgeworth, and others. Osmaston after his 13 years of collection published "A Forest Flora Of Kumaon", which included shrubs and trees while trekking from Srinagar to "Valley of Flowers" prepared a list of plants. Rau also made comprehensive collections from the interior parts of North Garhwal followed by Naithani and Singh & Singh who explored the adjacent localities of Garhwal Himalaya.

A perusal of literature indicates that earlier explores paid little attention to this zone of Himalaya. The present exploration was carried out with a view to compile a list of plants, as complete as possible, that exist at Gopeshwar and nearby places. The study would be helpful to the scholars working on various aspects of vegetation from this interior part of the Himalayan basin.

The area under study includes Gopeshwar and adjacent places of Dasholi Block, ranging from 700 to 4,200m of altitude, lying in the district of Chamoli Garhwal. The town of Gopeshwar is situated at an altitude of 1,300 to 1,500m on intermediate hilly ridges which continue higher towards North East and lie between 30°N latitude and 79—25′E longitude. It is encircled by high hilly ranges followed with slopes and the valleys. The area is delimited by river Alaknanda and its tributory Balkhila in the South East and North West, respectively.

All villages and the town of Chamoli, comprising the area under study, are situated up to 2,000m above sea level. In the north of Gopeshwar there are famous shrines and temples of Tunganath, Ansuya, Chandrashila Peak and villages of Mandal, Bairagana, Sagar, Gangolagaon, while religious shrines of Rudranath and village namely Rauli, Dewar, Khadora are on its North East. Snowy peaks of Nanda Ghunghati are visible marginally being situated far away after some hilly ranges in the East. Along the banks of Alaknada river in the South there are villages such as Paduli, Kothiyalasin, Chamnau, Patiyaldhar, Balkhila, Maithana and a small township, named Chamoli (formerly known as Lalsanga) which is on the main road leading to Badrinath.

In the West there are Gair, Tangsa and Bamiyala villages of the Balkhila valley. With the lapse of time, changing course of water streams, lanslides, floods, vegetational degradation, other biotic factors and electric thundering, especially at higher elevations, have caused variations in general topography of mountains and vegetation. Valleys with cliffs and gorges have all sort of exposure and faces.

Basic types of rocks present in this area re granite, dolerite, epidiorite, and igneous. Sedimentary rocks consist of limestone and stone, conglomerate, breccia, shale, etc. wheel quartzite, dolomite, shlorite and mic aschists are abundant in the metamorphic rocks. Soil is mostly sandy along the back of Alkananda and to some extent in other ravine valleys. It varies from alkaline sandy loam to clayey loam in open agricultural fields of low lying areas while there is humus rich soil in oak forest, acidic in pine forest and gravel to chert type in alpine region. Alkalinity, acidity and mineral nutrients also vary from place to place depending upon the nature of parent rock and its units.

Several ecological factors are responsible for a unique feature of the climate, varying from sub-tropical to alpine with the usual rainy winter and summer seasons. Frequent rains occur at high elevations but the total amount of precipitation decreases with increasing elevation. Maximum rainfall occurs during rainy months of July and August. Heavy and early snowfall along with frequent rains are common in alpine zone with gradual decrease and late occurrence common in alpine zone with gradual decrease and late occurrence down to 1, 500 m of altitude during winter months of November to February. Temperature in valleys during this period ranging from 15° to 24°C, comes down to 0°C while there is a gradual upward trend from March onward and it reaches maximum during late May to early June.

Corresponding to the climate and topography of the area, there are sub-tropical, temperate and alpine types of vegetation ranging from 750-1, 600m, 1,600-2,900m and 2,900-4,200m of altitude, respectively. Different factors have influenced the normal limits of vegetation types and flowering seasons. Natural vegetation near towns and villages in the sub-tropical belt has been subjected to more biotic disturbances as compared to the villages at higher elevations. There is abundance of grasses and annuals during monsoon, whereas perennials, shrubs and trees mostly bloom during spring and summer. Subtropical zone comprises mainly xerophytic vegetation throughout the bank of Alakanandas and in the lower portion of Balkhila valley. It is represented by shrubs like Ziziphus mauritiana, Caesalpinia decapetala, Euphorbia royleana, xanthium strumarium, Opuntia vulgaris, Woodflordia fruticosa, Cryptolepid buchananii, Artemisis scoparia, Carissa spinarum, Adhatoda vasica and trees e.g. Mallotus philippinensis, Sapium insigne, Emblica officinalis, Rhammus triqueter, Bombax ceiba, Boehmeria rugulosa, and Phonex humilis.

Deep in the Balkhila valley, this vegetation is replaced by the plants of ravines and include trees of Alnus nepalensis, Spondias pinnata, Engelhardtia spicata, Pyrus pashia, etc. and characterostoc shrub such as Callicarpa macrophylla. Other shrubs are Plectranthus japonicus, Craniotome versicolor, Buddleja neemda, Rhamnus triqueter, Urtica dioica, Arundo donax, Artemisia nilagirica, Berberis asiatica, Rubus ellipticus, and Rosa brunonii. Some of them are also common along road side and on waste lands, and include climbers such as Galium, Clematis, Ipomoea, Dioscorea, Cissampelos, Stephania, and Rubia. Some of the common herbs and weeds are Rumex hastatus, Dicliptera roxburghiana, Reinwardtia indica, Coleus forskohlii, Tridax procumbens, Cynoglossum spp., Strobilanthes spp., Polygala abyssinica, Crotalaria calycina, Seigesbeckia orientalis, Micromeria biflora, Thalictrum foliolosum, Boenninghausenia albiflora, common weeds Cannabis satica, Tagetes minuta, Chenopodium ambrosioides, Cirsium wallichii, Erigeron spp., Euphorbia spp., Euphorbia spp., an Bidens spp. etc. On steep metamorphic slopes, opposite to Alakananda and Balkhila and deep in the North of Gopeshwar, there are forests of Pinus roxburghii ranging from 1, 000-1, 600m of altitude.

Undergrowth of such foreses mostly consists of grasses and occasional shrubs. On other aspects of the slopes starting from valleys, grasses such as Arundinella nepalensis, Apluda mutica, Andropogon tristis, Cymbopogon distans, C. martinii, Oplismenus spp. Agrostis, Eragrostis, Bothriochloa, Panicum, Cynodon dactylon, Saccharum spontaneum and Cyperaceous members e.g. Carex condensata, Eriophrum comosum etc. constitute the grassy slopes. Associated herbs are species of Desmodium, Crotalaria, Galium Cassia, Swertia, Bupleurum, Phaseolus, Vicia, Striga, and others. Wider portion of the valleys and terraces of surroundings slopes of mountains are useful from physio-agronomic points of view. Crops are grown twice a year besides seasonal vegetables, fruits etc.

Important crops are of major cereals-wheat, paddy, barley, maize, millets, finger millet, Setaria italica, Panicum frumentaceum pulses-Pisum sativum, Phaseolus mungo, Glycine max, Lens culinaris, Cajanus cajan and oil crops like Brassica compestris, Sesamum indicum etc. Common weeds of winter season crops, which are locally known as Kuru, include species like Anagallis arvensis, Zornia gibbosa, Lathyrus aphaca, Fumaria indica, Stellaria media, Phalaris minor, Vicia hirsuta, Antirhinum orontium, Soliva anthemifolia, etc. Among weeds of rainy season crops, which are collectively known as Guderu or Challa, are Cyperus spp., Polygonum spp., Euphorbia spp., Chenopodium spp., Galinsoga parviflora, Oxalis corniculata, Panicum psilopodium Fimbristylis dichotoma, Echinochloa crusgalli, and others.

Near villages, trees are raised for fruits, fodder and other purposes e.g. Celtis caucasica, Grewia optiva, Juglans regia, Citrus spp., Toona serrata, Ficus spp., with occasional association fo Mangifera indica, Carica papaya, Ficus religiosa, F. bengalensis in the lower limits of the valley. Farms of sericulture include the cultivation of Morus alba and Quercus leucotrichophora to yield raw silk and tussar, respectively.

Many ornamental an shady trees have been introduced on the oradsides e.g. jacranda mimosaefolia, Bauhinia variegata, Cassia fistula, Eucalyptus spp., and a few others. With the end of upper limits of pine forests and ont he northern slopes above Gopeshwar, there starts temperate mixed vegetation. Forests are locally known as Patal. Major tree components are of Quercus leucotrichophora associated with Myrica esculenta, Rhododendron arboreum, Litsea umbrosa, Alnus nepalensis, Lyonia ovalifolia, Osmanthus, fragrans, Cotoneaster bacillaris, etc. Ground vegetation, in exposed forest pathces, consists of shrubby species e.g. Berberis chitria, Deutzia corymbosa, Elaeagnus umbellata, Pyracantha crenulata, Girardinia palmata, Princepid utilis, Rubus lensis. Among the common herbs Pedicularis spp., Plumbago zeylanica, Brunella vulgaris, Geranium spp., Primula denticulata, Gentiana spp., Potentilla spp., Myriactis nepalensis, Swertia chirata, Anaphalis margaritacea, Trifolium repens, Bergenia ciliata, Circium arvense and grasses like Erianthus rufipilus and Miscanthus nepalensis constitute the ground vegetation.

Shady cool aspects are characterised by trees like Quercus florilbunda, BEtula alnoides, Carpinus viminea, Eurya acuminata, shrubs i.e. Daphne papyracea, Sarcococa saligna, and Arundinaria spp. grasses Chrysopogon gryllus and Eulalia quadrinervis; andcyuperaceous species and herbs lije Taraxacum offiicinale, Tragopogon gracile, Habernaria pectinata, Salvia lanata, Androsace lanuginosa, Erigeron spp., and several others. The higher elevation limits of temperate vegetation are represented by trees like Quercus semicarpifolia, Syringa emodi, Abies pindrow, Texus wallichiana, Acer caesium, Betula utilis and the ground vegetation includes Leycesteria formosa, Salvia glutinosa, Polygonum spp., Ophiopogon intermedius, etc. Alpine vegetation is usually of herbaceous nature, locally known as Byghals with procumbent bushy patches of Rhododendron barbatum, etc. at lower extremities. Herbs with their limited span of beautiful flowering being to sprout as snow starts melting. They include species of Polygonum, Primula, Aconitum, Potentilla, Cynanthus, Orchis, Morina, Valeriana, Anaphalis, Inula, Saussurea, Senecio spp. Allium, Gentiana, Ranunculus, Thalictrum, Cotoneaster and abundance of soft grasses and ferns like Osmunda spp.

Flowers of Saussurea obvallata find a place in the temples of Rudranath as being offered during worship. Grazing is common during rainy season by sheep, goats, mules, buffalowes of Gujars and local inhabitants. While the genera and species are arranged alphabetically, the families are listed according to Bentham and Hooker's system of classification with slight modification as per modern concept. An attempt has been made to incorporate the latest change sin botanical nomenclature of plants, Local/vernacular names of some plants have been given within brackets following the botanical names, after verification from the local inhabitants to help in the future identification.

Each species is followed by a brief not about its habit, colour of flower, nature of inflorescence, peak flowering time, habitat, approximate altitude or range of occurrence and the corresponding herbarium number.

Kumaon Himalaya is perhaps the most attractive regions from the scenic as well as the floristic point of view among all the sectors of the Western Himalayas. It is located between 28° 43° to 30° 49°N latitudes and $78^{0}44'$ to $81^{0}4'$ E longitudes and includes three districts viz. Naini Tal, Almora, and Pithoragarh. The high altitude regions are chiefly in the district Pithoragarh especially in Dharchula and Munsiari tehsils except Pindari which fall sin Almora district. The important places from where plants have been collected earlier are Pindari and its morainic environs, Panchuli basin, Milam and Shelong Glaciers, Suryakund, Sangcha kind, Trisul basin, Brijikang pass, Ralam valley, Upper Dwali and Kali valley. Several other places have been botanised by the author during the past five years.

The prominent plants forming the general vegetation of alpine region are : Betula utilis D. Don, Rhododendron campanulatum D.Don., R.lepidotum Wall. ex G. Don., R. anthopogon D. Don.Lonicera angustifolia Wall. ex DC., L. obovalata Royle ex hook.f. & Thoms., L. purpurasens Hook.f. & Thoms., L. webbiana Wall. ex Lindl., C. acuminatus Lindl., Ribes glaciale Wall., R. emodense Rehder, Berberis umberllata Wall. ex G. don., B.kumanonensis Schneid., Rosa sericea Lindl., R. macrophylla Lindl., Caranga gerardiana Royle., C. pygmaea D., Hippophae rhamnoides Linn., Myricaria elegans Royle, M.germanica Desv., Salix fruticulosa Anderss., S. sclerophylla Andress., S. wallichiana Anderss., and several others. These are conspicuous woody representatives of the alpine flora and the herbaceous elements are represented by species of Astragalus, Corydalis, Polygonus, potentilla, Saxifraga, Gentiana, Impatients, Sedum., Pedicularis, Senecio, Primula, Epilobium,/saussurea, Swertia, Rumex, sedges and grasses. Plants have been seen to grow generally up to 5,600m.

The alpine zone usually begins where the tree zone ends. This is known as the timber line. A fringe of shrubs, scattered or continuous dwarf or gnarled trees, is found immediately above the tree line. These stunted trees exhibit a characteristic wind-wept or snow-wept appearance because of their exposure to the fury of cold winds, blizzards and snow storms. Birches, rhododendrons and willows among others are usually seen in such places.

The high altitude conifers, e.g. species of Abies, Pecea, Cedrus, and Pinus are seen associated with broad-leaved trees like Quercus semicarpifolia and Betula utilis at the altitude of about 3,600m. B. utilis may often be seen forming pure stands or usually forms association with Rhododendron campanulatum. Among the shrubs found above the tree line are usually species of Berberis, Myricaria. Lonicera, Rhododendron, Salix, and Jumpierus. While going on excursion in high altitudes, one usually finds a grassy belt with lush meadows, grassy slopes and flower-laden pastures along with several prostrate shrubs like willows and heathers above the tree line. These meadows and grassy slopes offer an important place for alpine plants and are well-known for their beautiful and attractive flowers. In late May and early June, the summer flowering herbs being to flower in great profusion. Among them are several crucifers, primulas, anemones and saxifragas and late in summer and towards autumn appear the gentians, swertias and others with saussureas attaining their best development late in autumn.

In Kumaon, the high altitude pastures and grassy slopes are locally called as bugyals. These are generally found at altitudes above 3, 500m and several beautiful and graceful herbs occur in great profusion on the gentle slopes in the bugyals. The bugyals generally show mixed association of grasses, sedges and colourful herbs, belonging to various families but sometimes a pure association of grasses is also seen. These grassy meadows from a lush carpet in many localities particularly in Upper Gori valley. The distribution pattern of bugyal flora is largely governed by the local edaphic and climatic conditions.

The varied, abundant and colourful high altitude flora of Kumaon Himalaya has been explored by several professional as well as amateur enthusiasts. Among the persons who had involved themselves in this great work include administrators, army officers, missionaries, surgeons, and survey officers who contributed towards most for the advancement of present day knowledge of flora. William Moorcraft, Trail, Lindsay, Edward Madden were the collectors in this area during the early years. The most extensive and the best collections in Kumaon was that of two survey officers, Strachey and Winterbottom. A list of which was later published by Duthie. George King, Duthie and his collectors, Wallich, Falconer, Royle, Drummond, etc. explored this area towards the end of the past century.

Two important publications on the orchids of North Western Himalaya and a revised list of plants of Kumaon and Garhwal, were authored by Duthie. Osmaston brought out his 'A Forest Flora for Kumaon'. He described and enumerated only trees, shrubs and climbers but did not include herbaceous vegetation. Several other workers have studied the Apline flora of Kumaon.

Analysis of present day alpine flora of Kumaon Himalaya reveals 662 species belonging to 251 genera and 57 families. The number of genera and species have been given in parentheses for each family. The families represented in this zone are: Compositae (27/80), Ranunculaceae (16/39), Cruciferae (23/37), Rosaceae (9/36), Papilionaceae (12/34), Gramineae (18/32), Saxifragaceae (4/30), Caryophyllaceae (9/29), Scrophulariaceae (7/28), Primulaceae (3.22), Cyperaceae (4/21), Umberlliferae (12/20), Gentianaceae (4/19), Polygonaceae (4/19), Liliaceae (12/18), Lamiacace (10/16), Fumariaceae (1/14), Caprifoliaceae (3/14), Crassulaceae (2/13), Juncaceae (2/12), Onagraceae (1/10),

Orchidaceae (5/9), Campanulaceae (3/9), Ericaceae (3/6), Berberidaceae (2/5), Valerianaceae (2/5), Papaveraceae (2/4), Geraniaceae (1/4), Balsaminaceae (1/4), Rubiaceae (1/4), Oleaceae (2/3), Solanaceae (3/3), Dipsacaceae (3/3), Orobanchaceae (1/3), Urticaceae (2/3), Violaceae (1/2), Tamaricaceae (1/2). However families like Araliaceae, Betulaceae, Cannabinaceae, Convolvulaceae, Fagaceae, Iridaceae, Circaeasteraceae, Asclepiadaceae, Lentibulariaceae, Loranthaceae, Polemoniaceae, Potamogetonaceae, and Zannichelliaceae merely have only 1 genus and 1 species.

Some of the plants like Thlaspi andersonii, Chorispora sabulosa, Christolea himalayensis, Creamanthodium nanum and others have a restricted distribution being found only in Kumaon Himalaya and in Western Himalaya or in adjacent Tibet. There are species like Thalictrum alpinum, Ranunculus hyperboreus, Potentilla multifida, Oxyria didyma and several others which enjoy a world-wide distribution in the alpine as well as arctic localities.

Cicer soongaricum and Lamium rhomboideum extend north wards to the Central Asian high-lands. Gueldenstaedtia himalaica, Saussurea leontodontoides, Cicerbita macrorhiza, Picrorhiza kurrooa along with others are distributed along the Himalaya from west to east. Alectris pauciflora and Anemone repicola are found in West China and also throughout himalayas.

5 Wasteland Management

The developmental activities in the Himalayas have to face the challenge of these geological hazards and in some cases the developmental activities, such as the impounding of large reservoirs of water, tunneling operations and the addition of silt load of rivers by road-cutting lead to the intensification of the natural dynamic process. Despite their critical importance, over 40 m ha in Indian Himalayas have been seriously degraded. Increasing population pressures and destructive land-use have resulted in the loss of fuelwood and fodder supplies, greater flood damage, intensification of drought, sedimentation of dams and reservoirs, and loss of crops and livestock. This has caused unnecessary poverty in the hills and unnecessary damage to the low lands. More than 270 million people in India are hut by this process. These resources reflect the importance of the Himalayas in Indian context.

Though the fragility of the Himalayas is well-known, the developmental activities did not take into account this aspect. The vast Himalayan region that extends from Jammu and Kashmir in the west to Arunachal Pradesh in the east has been under severe deforestation due to various developmental activities. The exposed land is swept away by rains causing floods in the rivers and sedimentation in reservoirs. This has created the problem of siltation in the dams thereby reducing their capacity as well as effective age. The areas which are directly fed by the Himalayan rivers and which form the Gangetic plains, the sediment coming from the Himalayas.

The damage in the form of floods has been quite catastrophic. On an average, floods annually affected an area of about 8.8 m ha of the Indo-Gangetic and Brahmaputra plain, including some 3.5 m ha of cropped land. Deforestation caused by the need for domestic fuel and timber supplies and livestock fodder, overgrazing of natural grasslands, indiscriminate cultivation on steep slopes combined with poor agricultural practices and badly designed roads, all contribute to the general ecological decline observed in many parts of the Himalayas.

The extent of first destruction that has occurred over the years has varied in different states depending on many factors, but it is clear that the rate of destruction is much more than the rate of afforestation. The outcome of these activities is increasing soil erosion which threatens the productive capacity of habitable areas of the Himalayan range and at the same time increase flash flooding. Principles of Strategy for the Himalayas

(a) Relieve pressure on the damaged ecosystem.

(b) Satisfy the demands of the people (and cattle) who exert this pressure.

(c) Use technology to satisfy the above needs.

(d) Control and management by the people of the series of projects (programme approach).

(e) Organise and train the people for task number III and IV-People to take responsibility.

(f) The project activities be based on.

(i) Cash crops

(ii) Tree crops

(iii) Side line occupation.

In order to provide short term, medium term and long term cash and other benefits to the participants of the programme six rules of technology can be adopted:

(a) Stop soil erosion along the road-side.
(b) Use land along the roadsides for "Protein Path" planting of laguminous.
(c) Land preparation and Plantation on the hill slopes.
(d) Find and use water efficiency.
(e) Promote protein banks for cattle.
(f) Landslide stabilisation.

In the call for action, seven pronged action programme which in brief asks for increased tree plantation, decreased deforestation, improving springs and water harvesting, promotion of renewable sources of energy, integrated programme fro improving crops, livestock, agro-forestry, grassland, bio-fencing, and intensification of poverty alleviation programmes, etc.

The strategy also includes identification of critical catchments and their repair, as also an urgent call to the people and to the government for a concerted action. The proposals for accelerated action identify:

(i) The problem.
(ii) The success stories.
(iii) Proposals for accelerated action.
(iv) Recommendation for Five Programme (1987-91) including investment needs.

The "State of Himalayas 1986-A Call for Action" has been prepared by Dr. M.L. Dewan, Member (Soil Conservation) national land Use and Conservation Board, on invitation of Ministry of Agriculture-Land Resources

Commissioner. The emphasis is on people's participation, first report of which was published in 1985. This followed with several programmes-eco development camps, nurseries, plantation, formation of local conservation and forest societies in the Himalayas.

First and foremost, it has to be accepted by all concerned that the best and the most practical route for conserving natural resources (soil, water, vegetation, etc.) is through their wise and rational utilisation. The local communities place a value on the resources; it is precisely because of this that they are utilising them and are benefited by the use. What is important is not to alineate the local communities by dropping a "legal curtain" between the people and the resources but to find ways and means of seeking people's participation for sustainable use of the resources. This plan, therefore, places the people at the centre of the stage and calls for management of the resources through them and by them. The objectives of the Call for Action are :

(a) Through people's participation achieve :
   (i) the protection of the soil;
   (ii) the promotion of employment of local communities who should be the partners in conserving and utilising the natural resources in a sustainable manner;
   (iii) The acceleration of income and the improvement in the living standards of the people;
   (iv) development and improvement of the availability of surface as well as groundwater resources;
   (v) promotion of viable and sustainable framing systems which are integrated in nature and serve not only the comprehensive needs of the farming communities but are also environ-mentally sound.

(b) To ensure that floods are prevented thus contributing to the development of sustainable agriculture in the lands. The reduction of siltation of the rivers has manifold other benefits.

(c) Conservation of the uplands and the catchments will prolong the life of the major and minor reservoirs that are already constructed.

In view of scarce resources, the focus should at least for the coming five-year period be on the most severely degraded areas which are also most densely populated. A survey is being undertaken (using satellite imagery and aerial photography) which aims at identifying the specific villages and the lands around them that need to be developed. Many of these catchments are identified.

The NGOs and voluntary agencies undertake leadership in each catchments, tasks proposed and 7 pronged action plan outlined. After detailed action plan, the task is to act. Critical watershed are being selected for urgent and concerted action. Non-government and people's organisations will give greater attention to these critical watersheds which are identified as giving greater damage in flooding, erosion and siltation.

Government can continue its plan but in whatever way it can help the local societies it will be extra effort. These critical catchments are based on detailed soil surveys as well as use of remote sensing imagery and aerial photographs. These are the critical catchments which are defined as "where the quantitative studies have shown high values and needing urgent attention due to high erosion evidence by the following data."

(i) Sediment yield index weightage values use (potential soil detachment).

(ii) Delivery ratio representing percentage of silt detached reaching the reservoir site (or potential reservoir site).

(iii) High slope percentage, shallow depth of soil, calcareousness, gravelliness, soil cover, land surface characteristics, existing erosion conditions such as gully, ravine and landslide conditions, etc.

Some of these critical catchments need more studies before delineating them exactly on the map and on the ground. The "State of Himalayas 1986: A call for Action" is based on seven pronged strategy as follows:

1. Increase production and planting of trees
   (a) Regeneration.
   (b) Planting programmes of fuel, fodder fruit and multi-purpose trees.
   (c) Bio-fencing/social fencing.
2. Reduce consumption of woodfuel and cutting of forests
   (a) Alternative energy (solar/biogas/wind, etc.,)
   (b) Mini hydel.
   (c) Smokeless chulhas.
   (d) Other measures for energy conservation.
3. Special treatment of critically damaged areas
   (a) Select critical micro/mini catchments.
   (b) Treatment around villages/towns, where damage is extensive.
   (c) Landslides must be treated by special means (In situ planting Juniper, other evergreen, etc.)
4. Revival/regeneration of springs and ponds/water harvested in small scale water dams+mini micro hydel + not large dams.
5. Promote integrated system with agro-forestry horticulture crops, livestock, grasses, fish and related tasks based on land capability and increased employment and moisture of people using. Terraces agro-forestry and water arrangements and related

systems for optimum land-use regarding agro-forestry is past of this system.

6. A Call to People

   (a) Family planning is most urgent. Produce less children.

   (b) Take the challenge. Be self-reliant.

   (c) Establish local conservation societies, for survival work cooperatively and not always depend on the government, district forest association, people's nurseries, catchment development board.

   (d) Organise into special sectors-special groups such as:

      (i) Ex-servicemen.

      (ii) Students.

      (iii)NGO/voluntary groups

      (iv)Women.

      (v) Other youth.

7. A Call to the Government

(a) Help those below poverty lines.

(b) Promote integrated farming.

(c) Utilise wealth of NGOs, voluntary organisations; military, paramilitary, border road organisations, ex-servicemen.

(d) promote people's and school nurseries.

(e) Assist in creating catchment development boards, district forest associations, village van panchayats.

(f) Undertake training programmes in all aspects of Himalayan eco-development.

(g) Speed up the creation of a research, development and training centre in the Himalayas like the one proposed Centre for Himalayan Conservation and Development.

(h) Decentralise decision making process.

(i) Have a Central Coordination Cell to monitor and evaluate the ongoing programme in the Himalaya.

Development including ecological development present unusual challenge to both practitioners and theoreticians of development is developing man, developing people and not only things. This is the fundamental premise on which both ends and means must stand totally reconciled.

Development appears less and less to depend either on resources or on technologies by themselves, attention focuses more and more on people and institutions which can mobilize and manage resources and extend and adapt technologies. The essence of development is to raise productivity which is achieved through an integrated action of land, labour and capital and is defined as economic development. What the economist call people. Technologies, resources and people's institutions each contribute in different but complementary ways to raise productivity.

Strategies and concerted action are best executed if the people who participate in such execution understand and see their own interest both short-and long-term in such development. Focus should be on strengthening of local capacities for planning and implementing development activities. Analytical framework should define the three dimensions of participation and collective action. Who: who participates; what; In what kind of participation; and how: what are the quantitative aspects-voluntary or coercive participation-continuous or *ad hoc*.

The dimensions can be described as ranging from highly selfish to highly altruistic and from highly individualistic to highly cooperative. These dimensions define a "space" within people's "values" can be located.

The concept of "social energy" to account for the dynamism observed in certain grass-root projects (example Dasholi Gram Panchayat Samiti, Gopeshwar, Chamoli District. U.P. - Role of Stimulator such as Chandi Prasad Bhat). A concept like "energisation" although not quantifiable nor predictable, can still have real and significant consequences. When they are achieved. They are remarkably cost effective.

Possibility of "free-riding" will inhibit collective action in general and the individual acts of "piracy" will inhibit regimes of natural resources management such as soil and water conservation in the Himalayas. Example can be 'political interference' and 'political free riding' or 'piracy' at the local level and such examples are plenty in and outside the Himalayas. The "tragedy of the commons" is created by persons who overuse a 'common property' resource like forest or pastural land to others disadvantages even to the point of destroying the resources. This is what is happening in most parts of the Himalayas.

Several Governments, have failed to manage common resources without full participation of the local institutions mainly because of bureaucratic approach and distance from the site as well as inadequate interest. Realization of the approach by the farmers and rural people of self-reliance because they learnt that if they did not help themselves, nobody else would. Development of organisational capacity for farmers to participate in the management of common resources. Such development should ensure for farmers a continuous role in the planning, operating and maintenance of the common system so that it would not lapse back in its sad and unproductive condition.

The creation of local institutional capacity to manage natural and other resources through collective action, not dependent entirely on the structures and authority of the state offers an important alternative to bureaucratic approaches, if such capacity can become viable.

One of the reasons for ambiguity and confusion in such of the thinking about "local" is that "local" has both a social and geographic meaning and neither may match the administrative definition given to the term. In words of Norman Uphoff renowned social scientist, it is useful to recognise that there are at least ten levels of decision-making and activity which range from the international (supernational) level down to the level of individual decision makers. The levels (h) to (j) of these come within the range of what is usually meant by "local".

(a) International Level (e.g.) donor agencies, IBRD, FAO, UNDP, UNICEF, etc).

(b) National level (Central Government agencies more than one level).

(c) Regional level (e.g. State level).

(d) District level (district administration).

(e) Sub-district level (Taluka, Tehsil, etc).

(f) Locality level (a collectivity of communities often centered around a market town).

(g) Community level (so-called "Natural Village").

(h) Group level (neighbourhood residential group, occupational group, caste or other collectivity).

(i) Household level.

(j) Individual level.

The first two levels (international and national) are far removed from "local" conditions and activities. The next three levels (regional, district and sub-district) are closer but still not readily local in any sociological sense, but being essentially administrative entities. There is no basis or capacity at these levels for organizing activity to promote development, except through formal political channels or through diffused market relationship. Only when we get to a set of communities which have some established patterns of communication, economic exchange and social cooperation usually corresponding to what is often called "market town" or "block development" there is a potential for collective action which can be enlisted to carry out some of the responsibilities and burdens otherwise handled by the national government.

People within a locality have some acquaintance with each other directly or indirectly, through intermarriage, trading relations, religious festivals or participation in local government activities. This establishes some common identity which makes collective efforts more possible.

In the search for more effective approaches to involving the people and specially the rural people in their own development of their environment and ecology)eco-development) attention has been focused almost exclusively on the intended beneficiaries/participants. There is a familiar tendency to "blame the intended beneficiary". Failures to establish or maintain local organisation and to get individual cooperation are usually attributed to the people, their programme should assist their attitudes, their poverty, their culture, their literacy, etc.

Too often overlooked is the reality that the response of the poor to government progress is shaped by the way the services are administered:

(a) How accommodating or inflexible the services are?

(b) How satisfying or humiliating their treatment is?

(c) How readily the poor get access to services or how much more readily the rich can utilise them?

(d) Whether government staff adopt a problem-solving stance or a conventionally bureaucratic one?

(e) How attuned staff are to the actual conditions and needs of the poor?

(f) Whether these staff deal with the poor as responsible adults or as basically ignorant and irresponsible?

To expect rural poors' participation is quite unrealistic when the style of programme administration curbs their initiative, problem-solving behaviour and development of increased capacity for rural resource mobilisation and management at the local level-all essential elements of an authentic development process.

Where programmes are viewed by professionals as solutions to presumed beneficiary problems, any effort on the part of beneficiaries to engage in independent problem solving may even be construed as uncooperative or obstructionist behaviour. Through government staff may have good intentions, too often a programme in which there is no real participation by rural poor leads to delay, to inappropriate services and to apparently arbitrary demands by the programme upon them if they enter it. People are likely to be treated in patronizing ways by programme staff and programme itself may be misconceived and harmful to poor households struggling to survive in economic social and physical environments that are complex and fragile.

Recognising that development programme can and must benefit from ideas and resource inputs from the participants and the people themselves such programme commonly call for some form of local organisation-a society for local eco-development, an association for improved water use, a club for youth or mothers or women, etc. Even such organisation are created, it is possible that they may be created by the project people themselves and we of participation taking root, Often enough, the rural people such societies or associations. aucracies working with the people rural people, the reasons for why the pation remains stunted are not far to arge educational and status gap between aff and the poor.

there are ethnic or linguistic differences ng communication and mutual regard. There is on attitude of superiority which exalts 'expert' dge and deprecates what the poor know and are ble of doing. Even well-intentioned staff tend to fall the paternalistic postures which depreciate the pabilities of the poor or perpetuate the dependency syndrome. The participating people and more especially the rural poor cannot be expected to change their behaviour and their attitude in response to government programme unless and until government staff change their activities and attitude vis-a-vis the participating people and the rural poor. Blaming the participants or the rural poor should be avoided in all respects.

The bureaucratic re-organisation must involve individual attitude and value changes, the more important part involves changes in job definition. Performance criteria, career incentives, bureaucratic procedures, organisational responsibilities and the like. The problem of bureaucratic re-organisation has not been addressed in a systematic way more especially in bureaucratic organisation in rural and eco-development. Basically, the poor performance is largely as a result of a disparity between organisation's internal structures and operating systems on the one hand and requirements on the other of alleviating poverty and creating self-sustained, self-reliant development efforts. A strong

management would not be successful unless t[h]
better fit between ends and means. Apparently, t[h]
need to incorporate into existing traini[n]
development professionals certain modules
develop such skills in reorientation of bureaucrac[y]
requirements of fitting the task to the need.

There is also a need for a type of research
consulting services to achieve bureaucra[tic]
reorganisation. There are some inertial values which ha[ve]
rationalised bureaucratic convenience and advantage a[n]
seldom questioned by administrators. These values an[d]
presumptions are deeply institutionalised in bureaucratic
operations and routines. Some of these inertial values are
subject to question in the development process. They are:

1. *Expenditure-Results* : For decades, prevailing economic theory has equated development with growth of GNP. This is a mistaken and misleading formulation. Moving money does not necessarily mean development.
2. *Education-Superiority*: This also assumes that the judgement and even the values of educated persons should prevail. This assumption truncates participatory process. In fact, the formal knowledge of well educated persons is usually necessary but not sufficient for programme planning implementation. There is much local knowledge which the expert is unlikely and sometimes even unable to know and which local people can and should contribute to decision making processes. Local knowledge itself is unlikely to be sufficient, but it, is almost always necessary for efficient and effective use of development resources especially when the objective is to assist the poor majority. One of the purposes of getting more popular participation in development is to achieve a fruitful

combination of expert and indigenous knowledge.

3. *Projects-Development*: It has been the practice in development management to draw a sharp distinction between operating programmes and development projects. Within these framework, development is approached through a series of finite, discontinuous actions with discreet time-bounded outcomes which depend on special temporary injections of external funds. Projects implemented by temporary organisations to achieve limited time-bounded results are commonly defended as having positive demonstration effects, assuming that the new concept they demonstrate will be adopted by more permanent organisations once proven in the field. Experience does not necessarily hold such an assumption. Constant and widespread eco and rural development depends more on moving whole of section of the bureaucracy towards new mode of operations than on the enclave progress too often associated with projects. It calls for a strategic style of management which sets directions, creates new visions of the possible and builds new organisational capabilities for appropriate response to changing conditions. It is commonly observed that lack of commitment and imagination among the personnel of development agencies is more a consequence of bureaucratic systems which treat creative behaviour as disfunctional than a reflection of any inherent qualities of such personnel. The bureaucratic reorganisation can be approached through a learning process which has certain key points: (1) learning to be effective; (2) learning to be efficient; and (3) learnings to expand.

There are some elements for bureaucratic reorganisation which can be summarised as: (1) Strategic management;

(Reward Structure; (3) Planning systems; (4) Monitoring and evaluation; (5) Applied social science; (6) Personnel procedures; (7) Financial management; (8) Differentiated structure; (9) Training; and (10) Doctrine. All aspects are important.

As far as doctrine is concerned it determines the degree of commitment and devotion and are based on clear and widely shared understandings throughout the organisation about its mission, how it can and should assist the poor majority and how staff can and should be involved in this process. A framework for action in bureaucratic reorganisation needs a four-fold effort: (1) Theory and method development; (2) Dissemination; (3) Experimentation through formation of collaborative linkages between operating agencies and knowledge resource institutions; and (4) Skill building by training of interested professionals in the particular skills required to manage and facilitate bureaucratic reorganisation.

INDEX